W9-CWV-736

TECHNOLOGICAL CHANGE AND THE DYNAMICS OF INDUSTRIES
Theoretical Issues and Empirical Evidence from Dutch Manufacturing

CONTRIBUTIONS TO ECONOMIC ANALYSIS

253

ELSEVIER
Amsterdam – Boston – London – New York – Oxford – Paris – San Diego
San Francisco – Singapore – Sydney – Tokyo

TECHNOLOGICAL CHANGE AND THE DYNAMICS OF INDUSTRIES

Theoretical Issues and Empirical Evidence from Dutch Manufacturing

Machiel Van Dijk
CPB Netherlands Bureau for Economic Policy Analysis,
P.O. Box 80510,
2508 GM DEN HAAG

2002

ELSEVIER
Amsterdam – Boston – London – New York – Oxford – Paris – San Diego
San Francisco – Singapore – Sydney – Tokyo

ELSEVIER SCIENCE B.V.
Sara Burgerhartstraat 25
P.O. Box 211, 1000 AE Amsterdam, The Netherlands

First edition 2002

Library of Congress Cataloging in Publication Data
A catalog record from the Library of Congress has been applied for.

ISBN: 0-444-51177-6
ISSN: 0573-8555(Series)

⊗The paper used in this publication meets the requirements of ANSI/NISO Z39.48-1992 (Permanence of Paper).

Printed in the Netherlands.

INTRODUCTION TO THE SERIES

This series consists of a number of hitherto unpublished studies, which are introduced by the editors in the belief that they represent fresh contributions to economic science.
The term 'economic analysis' as used in the title of the series has been adopted because it covers both the activities of the theoretical economist and the research worker.
Although the analytical methods used by the various contributors are not the same, they are nevertheless conditioned by the common origin of their studies, namely theoretical problems encountered in practical research. Since for this reason, business cycle research and national accounting, research work on behalf of economic policy, and problems of planning are the main sources of the subjects dealt with, they necessarily determine the manner of approach adopted by the authors. Their methods tend to be 'practical' in the sense of not being too far remote from application to actual economic conditions. In addition they are quantitative.
It is the hope of the editors that the publication of these studies will help to stimulate the exchange of scientific information and to reinforce international cooperation in the field of economics.

The Editors

CONTENTS

1. INTRODUCTION

Industrial landscapes exhibit high degrees of diversity. In all developed countries, we observe a substantial cross-sectional variety in the ways in which the activities of producers are organised in meeting the demands of their customers. Consider, for example, the large variances in sectoral structures. Whereas some industries accommodate only a small number of large enterprises that are hardly ever challenged by entering firms, other sectors are much more characterised by a large population of small firms that continuously rejuvenates itself through the process of entry and exit. Interestingly, this observation goes together with some remarkable similarities between countries. Chemicals and oil refinery are typical examples of industries dominated by a few large firms, whereas sectors such as textiles and wooden products show much lower levels of sales concentration in most of the industrialised countries.

Besides the large diversity of the industrial landscape, we also observe that this landscape is subject to substantial change over time. The rise and fall of some industries and, within these industries, the evolving patterns of entry, exit, and growth and decline of firms continuously alter the industrial landscape at the various levels of analysis. For instance, in the United States the automobile industry initially experienced a rapid increase in the number of producers, but while market sales were still increasing, a sharp reduction in the number of firms took place. Nowadays, the number of firms in this industry has more or less stabilised at a low level, which naturally implies rather high concentration levels.

Traditionally, such high concentration levels have been among the main concerns of anti-trust policies. A high concentration of sales in an industry is usually associated with substantial market power of the firms in the industry and, consequently, with a suboptimal economic performance (in terms of welfare). Many economic policies therefore reflect the idea that having an industry structure characterised by the presence of many firms is the best way to minimise the market power of individual firms and hence to maximise economic performance. Obviously, the justification of these policies is strongly based on the following two hypotheses: (1) high concentration levels imply low levels of competition, and (2) maximising competition guarantees optimal economic performance.

From a theoretical point of view, these two hypotheses can be seriously questioned. Concentration levels do not necessarily indicate the intensity of industrial competition. Although we may observe a small number of large firms in an industry, competition between them, or the competitive pressures from potential entrants, may still be considerable. Even if the concentration level of an industry is found to be high for many subsequent years, still a large share of the market may have been transferred within the group of continuing firms or from these continuing firms to entrants. Finally, one could even argue that increasing competitive pressures might force inefficient firms to exit the market, leading to an increase in concentration levels.

The second hypothesis asserts that maximal competition assures optimal economic performance. Although this may perhaps maximise the economic performance of an industry in the short run, one may question whether such highly competitive markets maximise economic performance in the long run as well, as fierce competition may hamper technological progress. High levels of competition could adversely effect firms' endeavours to be innovative in two ways. First, fierce competition may suppress profits so much that firms cannot finance their investments in research and development from their retained earnings. Second, the profits resulting from successful innovation may quickly be eroded in highly competitive markets. Consequently, if firms expect they cannot sufficiently reap the potential benefits from their innovations, they may, in the worst case, decide to refrain from investing in research and development at all.

In order to assess the validity of these arguments, and thus to evaluate the appropriateness of anti-trust policies, empirical research is obviously of utmost importance. Empirical analyses of, for instance, the relationships between concentration, competition and technological change may learn us more about the existence and significance of the trade-off between the short-run and long-run economic performance of an industry. Since the outcomes of such analyses could directly be used for welfare analysis and economic policy making, empirical research in industrial economics can already be fully justified on purely normative grounds.

On positive grounds, the legitimacy of empirical exercises in industrial economics is undisputed as well. As in many other scientific disciplines, theoretical and empirical work are strongly related in two ways. By exploring databases we may find empirical regularities that could provide a useful starting point for the identification of the phenomena to be explained. As such, empirical work contributes to the initial setting of a theoretical research agenda. Alternatively, by confronting theoretical conceptualisations with empirical data we are able to evaluate the explanatory powers of the models under investigation. If necessary, the output of such analyses can in turn be used to further improve, expand or redirect theoretical work in industrial economics.

In conclusion, on both normative and positive grounds, empirical research is essential for the analytical rigour and practical relevance of industrial economics as a scientific discipline. Naturally, sound empirical research in this field imposes strong requirements on the availability of economic data. However, for many decades industrial economists had to rely on case studies and on rather highly aggregated cross-sectional data. The availability of only this type of data has indeed affected the research agenda in this era. Since these data did not allow for empirical research on the long-run impact of, e.g., entry and exit, or the transfer of market shares within the group of continuing firms[1], not much effort was taken to develop or advance theoretical frameworks analysing the causes and consequences of these phenomena. Instead, theoretical work mainly concentrated on static equilibrium analyses of industry structures with the structure-conduct-performance framework, originally developed by Edward Mason and Joe Bain in the 1950s, as the dominant paradigm. In this approach, industrial competition is regarded as a static phenomenon, i.e., as a state of affairs, of which the intensity can simply be assessed by a set of structural attributes.

However, over the last fifteen years the quality and availability of economic data has substantially improved. Especially the increased availability of longitudinal firm-level databases has greatly enlarged the scope of empirical research in industrial economics. The availability of these data has made it possible for researchers to follow the characteristics of a large number of firms over time. This opportunity does not only allow for studying patterns of birth, growth and death of individual firms, but also for an analysis of the cumulative effect of entry, exit, and variations in sizes or market shares of continuing firms. Furthermore, it

[1]Case studies could, of course, allow for studying these matters. However, as Baldwin (1995) argues, their limited scope precludes the type of generalisations that social science demands.

has become possible to analyse differences between industries concerning these cumulative effects, and to investigate how industries change over time.

The increased availability of these longitudinal firm-level databases has indeed led to a large amount of empirical work taking advantage of the opportunities mentioned above. This work has resulted in a number of robust empirical results that are hard to reconcile with the assumptions and analytical results of the conventional static equilibrium theories. For instance, it has been observed that in virtually all industries entry and exit of firms occur simultaneously. This result is at odds with models in which the entry process is viewed as a mechanism by which the industry adjusts itself to changing exogenous conditions. A second example is the coexistence of firms of various sizes in rather narrowly defined industries. Many industries exhibit a fairly skewed and stable distribution of firm sizes, indicating that at any point in time large as well as small firms simultaneously inhabit the industry. This observation seemingly rejects the notion of the optimal firm size found in textbook models on industrial organisation. The final example concerns the persistence of asymmetric firm performances. The 'fact' that, e.g., profitability differentials between firms in the same industry are persistent over time suggest that profitability levels do not converge to some 'normal' equilibrium rate of return.

These empirical findings constitute an important reason for the increased interest in dynamic approaches that we observe in the recent theoretical literature on industrial economics. However, the idea of analysing industrial economics from a dynamic perspective is certainly not new. In the early days of this discipline, prominent economists such as Adam Smith, Alfred Marshall and Joseph Schumpeter started from a dynamic approach as well in theorising about the organisation of manufacturing industries. Intra-industry competition was clearly viewed by them as a dynamic process, i.e., a process causing continuous change, driven by the entry, exit and growth and decline of profit-seeking firms.

The increased scope of empirical research, together with the evidence it has brought forward, is certainly not the only reason for the upsurge of dynamic approaches in industrial economic theory. Especially in the last two decades, technological change is increasingly being acknowledged as an endogenous process driving industrial competition and economic growth. Since the industrial revolution, technological change has obviously always been an important economic phenomenon, but for many decades it has mostly been regarded as an exogenous factor in models of industrial economics. These models simply assume that the exogenously created knowledge related to any innovation flows around freely and hence does not effect competition between firms.

In reality, however, both bounded rationality and the tacit nature of technological knowledge prevent that new knowledge spills over rapidly, enabling innovative firms to enjoy at least temporarily higher (dis-equilibrium) profits. In fact, the opportunity to enjoy these profits provides the economic incentives to firms' search for technological improvements. Competition between firms is therefore largely driven by the *process* of technological change, continuously disturbing the economic status quo. Obviously, this rejects the notion of competition as a static concept underlying the mainstream equilibrium models. The inability to incorporate (technological) competition as an endogenous, perturbing process has therefore led to an increasing dissatisfaction with these models and, consequently, to an increasing and renewed interest in theoretical approaches in which the dynamics of industrial competition are endogenously driven by the process of technological change.

These theoretical approaches can roughly be categorised in two groups. The first one focuses on technological regimes. A technological regime can be defined as a particular combination of opportunity, appropriability, cumulativeness conditions and properties of the knowledge base that underlies the innovative and productive activities in an industry and explains how these activities are organised in an industry. The second group of theories focuses on product or industry life cycles. These theories start from the observation that most successful products go through a number of distinct stages over their lives, and aim to explain how the structural and dynamic properties of the industry co-evolve with these product life cycles.

Given the presence of these theoretical frameworks embodying the dynamic interaction between industrial competition and technological change, and given the increased availability of longitudinal firm-level data, it is actually quite surprising that until now the opportunity of testing these theories with the longitudinal micro-data has hardly been taken advantage of. Most of the empirical work using these data focussed on revealing some strong empirical regularities, but one of the main issues in industrial economics, explaining the observed cross-sectional differences, has hardly been addressed so far. This virtually unexploited opportunity constitutes the primary objective of this book: to explain how and, by using firm-level data on Dutch manufacturing, investigate empirically whether the technological regime framework and the product life cycle approach can explain the observed differences between industries with regard to their structural and dynamic properties.

Understanding whether and how technological regimes and product life cycles shape the structures and the dynamics of industries is important for a number of reasons. As mentioned already, the evidence derived from longitudinal databases suggests that existing conventional theories are in need of improvement. Probably the main reason why these equilibrium theories are hardly corroborated by the data is that they mostly neglect the process of technological change. Since both the technological regime framework and the product life cycle approach explicitly recognise technological change as the major determinant of the competitive process, empirical evidence supporting or rejecting these theories could at least indicate which directions to explore and which directions to discard in the further development of industrial economic theory.

From a policy point of view, understanding how technological regimes and product life cycles affect the competitive process is important as well. As we have argued before, the justification of economic policies rely substantially on what we know, both theoretically and empirically, about the competitive process within industries. As an example, consider again the concerns of anti-trust policies regarding high concentration levels. The assumption underlying these policies is that high levels of sales concentration indicate a lack of competition. However, both the technological regime framework and the product life cycle approach can provide certain conditions under which intense technological competition and oligopolistic market structures are naturally conjoined.

For instance, a technological regime with high appropriability and cumulativeness conditions and characterised by tacit and complex knowledge is much more conducive to technological competition driven by large and established firms than to a competitive process driven by small firms. But also the product life cycle approach depicts a specific evolutionary stage in which dynamic increasing returns to technological change becomes an important determinant of the competitive process. As these models show, the outcome of such a competitive process necessarily implies strongly increasing concentration levels. Although under these circumstances market structures will not correspond to the textbook conditions of perfect competition, the typical concerns of anti-trust policies regarding the adverse effects of high concentration levels on industrial competition may still be largely misplaced.

In conclusion, understanding whether and how technological regimes and product life cycles shape the structures and the dynamics of industries is of substantial importance to both the theory of industrial economics and the practical implications of this theory for policy. We believe therefore that the present book will provide a valuable contribution to the empirical foundations, as well as to our theoretical understanding and the practical relevance of industrial economics. In what follows, the structure of this book and the underlying research methodology will be presented.

Since the regularities provided by earlier empirical research using longitudinal databases are the starting point of this book, chapter 2 will investigate whether the previously obtained 'stylised facts'[2] are also observed in the Dutch manufacturing sector. As such, chapter 2 aims to serve three objectives. The first one is to describe the database we have used in this book. We will provide some general information regarding, e.g., the variables that are included, the number of firms and the period it captures, and present some descriptive statistics of the dataset. Further, we will pay attention to one specific limitation of the database, namely its observation threshold: only firms with at least twenty employees are included in the database. The second objective of chapter 2 is to provide a survey of the stylised facts that empirical work in industrial economics have provided so far. Finally, we will investigate whether these stylised facts are observed in the Dutch manufacturing sector as well.

Chapter 3 surveys the theoretical literature on industrial economics, and demonstrates how the selected theories may explain variances in the structural and dynamic properties of industries. The main focus of this chapter will be on the theoretical issues regarding the technological regime framework and the product life cycle approach, but chapter 3 will also include an overview of the equilibrium models. Although these models have dominated the literature on industrial economics for a considerable time, we will demonstrate that the equilibrium approaches and their related assumptions inherently involve some theoretical and empirical limitations. We will then argue that the technological regime framework and the product life cycle approach are conceptually much more appropriate in providing plausible explanations for the structures and dynamics of industries, as they both embody elements such as bounded rationality and technological uncertainty that are close to empirical substance.

Finally, chapter 3 will demonstrate how the technological regime framework and the product life cycle approach may explain the observed

---

[2]Because the notion of stylised facts is perhaps slightly confusing, we would like to emphasise that using the terminology of stylised facts merely represents our way of characterising the process and outcome of creating broad characterisations of available patterns in the data.

structural and dynamic differences between industries. The basic arguments are as follows. Given different technological regimes, i.e., different combinations of opportunity, appropriability, cumulativeness conditions and properties of the knowledge base underlying the innovative and productive activities, the resulting different patterns of innovative activities are likely to affect the structural and dynamic properties of industries. Alternatively, theories and models on product life cycles explain and depict the evolution of an industry's structural and dynamic properties over its lifetime. Based on this approach, the observed cross-sectional differences may be explained by the different evolutionary stages that the industries occupy.

By using the longitudinal firm-level database on Dutch manufacturing, chapter 4 will investigate the extent to which the technological regime framework and the product life cycle approach can actually explain the cross-sectional variances in structures and dynamics. From both theoretical frameworks we will first derive a number of hypotheses. Next, we will classify the industries in the sample according to their underlying technological regime and the evolutionary stage they occupy. For the classification of the industries into regimes we will use a taxonomy of technology classes offered by Malerba et al. (1995). Ideally, some exogenous criteria would be used for the classification of industries into evolutionary stages as well. However, since these are not available, we will have to base our classification on the revealed patterns of some variables denoting the evolutionary stage of the industries in the sample. Based on these classifications, we will then test the hypotheses for the technological regime framework and the product life cycle approach individually. Finally, we will perform a number of regression analyses in this chapter to investigate the extent to which these approaches collectively account for differences between industries, and whether any interaction effects can be observed between them.

Although the technological regime framework and the industry life cycle approach provide plausible explanations for differences in the structures and dynamics of industries, we believe that both these theories still ignore a number of crucial elements. First of all, models on product life cycles generally focus on the emergence and evolution of only one product and its associated technology. However, in many industries we observe that firms repeatedly introduce or adopt new product technologies that replace the older ones. Second, both these approaches do not explicitly consider differences in the technological properties of the goods produced by the industries. Finally, in models on technological regimes and on industry life cycles the growth of a firm is generally determined by its relative (technological) performance. However, empirical studies on firm growth do not provide much evidence supporting such a relationship. Most of these studies suggest that the size of a firm generally follows a random walk with a declining positive drift.

In chapter 5 we will introduce a model on industry dynamics that attempts to include these three elements. As in Shy (1996), the degree of substitutability between the quality and the network size of a technology and the degree of compatibility of succeeding technologies are the key determinants of the simulation model presented here. However, Shy (1996) mainly limits his focus to the demand side, as he investigates how varying consumer preferences over technology advance and network size effects the timing and frequency of new technology adoption. Our focus in chapter 5 will be on the relation between the demand side and the supply side. Given variations in consumer preferences over quality and network sizes, and different degrees of compatibility between succeeding technologies, we will investigate whether the resulting differences in the timing and frequency of new technology adoptions affect the dynamics of the population of supplying firms. This would be an important result, because if the industrial properties are indeed related to the diffusion patterns, our model may provide an additional explanation for the observed structural and dynamic differences between industries.

Since we believe that it is of equal interest that the emergent properties of our model are close to empirical substance, we will also investigate in this chapter whether the results of our model are consistent with the stylised facts that empirical research in industrial economics has put forward so far. Finally, by varying a number of parameters reflecting the technological regime conditions, we will investigate whether the results of our model are significantly different under various technological regimes.

Chapter 6 concludes this book by summarising the main results and by presenting some concluding remarks. Further, we will suggest some directions for future research in this final chapter.

2. EMPIRICAL REGULARITIES IN DUTCH MANUFACTURING

2.1 INTRODUCTION

In the last fifteen years, the increased availability of micro-level data has led to numerous studies on firm and industry dynamics using longitudinal datasets. The possibility to follow a large number of individual firms over time allowed researchers to study the dynamics of industries without having to rely exclusively on case studies. As a result, the empirical foundations of industrial dynamics have been greatly enhanced at all three levels of aggregation. At the firm level, these data have provided evidence regarding patterns of birth, growth and death of individual firms and their potential determinants. At the industry level, we have been able to fully assess the impact of dynamic changes in industries by, for instance, measuring the cumulative effect of entry and exit, and the turbulence caused by variations in sizes and market shares of continuing firms. Finally, at the sectoral (i.e., manufacturing) level we have been able to directly compare different industries with respect to their dynamics, and investigate their structural evolution. Together, these empirical studies have led to a rich set of stylised facts in the field of industrial economics.

This collection of stylised facts is employed to construct the framework of this chapter in the following way. First, the presentation of the stylised facts aims to provide a survey of the most important results empirical literature in industrial economics has put forward so far. Second, testing for the presence of these stylised facts in the Dutch manufacturing sector learns us the similarities and differences between the manufacturing sector in the Netherlands and manufacturing sectors in other countries. Finally, it gives us the opportunity to provide a detailed description of the dataset that we use throughout this book. The next section starts with this.

2.2 THE DATA

Recently, Statistics Netherlands (SN) has made it possible for researchers to gain access to a firm-level longitudinal dataset. In this section we will describe the SN manufacturing database that we will use throughout this book. We will provide some general information about the database (e.g., the period it captures, the number of firms), and pay attention to the specific limitations of it, especially with regard to its truncation of certain types of observations.

The SN manufacturing database captures all firms with more than twenty employees (working 15 hours or more weekly) that have been active in the Dutch manufacturing sector between 1978 and 1992. In total, there are 10,246 firms in the dataset, of which 2,558 firms are present throughout the whole period. These continuing firms on average capture 53.5 and 52.4 percent of, respectively, total manufacturing employment and value of production. The remaining firms are only temporarily present in the dataset. Although these firms enter and exit the database, defining them as greenfield entrants and closedown exiters is not appropriate in most of the cases. The majority of the entering and exiting firms appears or disappears because of passing the threshold of twenty employees. For the period from 1986 and onwards, SN provides aggregated statistics of the specific reasons why firms respectively enter or exit the database. Over this period, on average only 8.9 percent and 21.1 percent of the firms entering and exiting the database were greenfield entrants or closedown exiters.

Unfortunately, the firm-level data we have access to do not allow us to distinguish the entry and exit categories in detail. We are only able to identify firms that enter and exit for other reasons than greenfield entry, closedown exit, and passing the observation threshold. Whenever firms enter of exit because of these 'other' reasons (such as mergers, acquisitions or administrative reasons) SN assigns an additional variable to these firms. However, this variable does not specify which of these reasons applies. Hence, we can only distinguish firms that enter or exit for 'normal' reasons (i.e., new firms that really enter, existing firms that really close down and firms passing the observation threshold) from firms that enter or exit for 'other' reasons (which is never real entry or real exit). To give an example: when two firms merge in a certain year, we observe in that year two firms exiting for an 'other' reason, and one firm entering for an 'other' reason in the following year.

Although most of the firms enter and exit the dataset because of passing the observation threshold, we will still label them as entrants and exiters.

We could have called them, e.g., pseudo-entrants and pseudo-exiters, but for convenience we have simply named them entrants and exiters. Certainly, any empirical test aiming to identify, for instance, the determinants of the entry and exit decision would be highly inappropriate when these groups of firms are considered. For cross-sectional purposes however, studying the impact of the entry and exit process still makes sense. If, for example, an industry is conducive to entrants we may well assume that it is less difficult for small firms to grow and pass the observation threshold, to stay above the threshold, and to obtain a significant market share in that industry compared to other industries. We believe therefore that for the purpose of comparing different industries the use of statistics measuring the impact of the entry and exit process is useful, despite the presence of the observation threshold.

Hence, we will use the following definitions throughout this book. An entrant is a firm present in the dataset for one or more years, but absent in 1978. Its year of entry is the year in which it is observed for the first time. If a firm is present for one or more years, but absent in 1992, we label it an exiter. Its year of exit is the year in which it is observed for the last time. Excluded from the groups of entering and exiting firms are firms that enter and exit the dataset because of mergers, acquisitions or administrative reasons. Finally, an incumbent or continuing firm is defined as a firm that is present in 1978 *and* in 1992.

Table 2.1 shows the variables that are annually available for each firm. Furthermore, for most observations we have information about the gross investments in fixed assets. In total, 17,584 observations (out of 79,662) have missing values on this variable, however. Finally, we used producer price index numbers at the 3-digit industry level to calculate the real values of the nominal variables.

In total there are 106 industries in the sample.[3] The original dataset also contained limited data on the petroleum industry (mainly refining), the optical and medical instruments industry and a number of miscellaneous industries. However these data did not cover the full period between 1978 and 1992. For compatibility reasons we have left these industries out of our analyses. Table 2.2 shows to what extent the remaining industries in the SN dataset cover the total manufacturing sector in the Netherlands. A list of all 106 industries can be found in Appendix 1.

---

[3]Because some small 4-digit industries were (technologically) so closely related, we have aggregated these industries into 3-digit industries, and in one case even into a 2-digit industry.

Table 2.1 List and Description of Variables Available in the SN Database

Variable	Description
Class of manufacturing	All firms are allocated to a 4-digit industry, according to Statistics Netherlands' standard classification of industries as of 1974. If a firm is active in more than one industry it is allocated to the industry from where it receives most of its revenues
Number of employees	The number of employees working more than 15 hours a week, employed by the end of September
Industrial sales	The revenues from selling goods manufactured in-house or by others, provided they are on the payroll of the firm
Total value of production	The sum of industrial sales, activated costs (i.e., costs made for the internal production of capital goods for internal use), changes in the stock of final products and other revenues
Total consumption value	The sum of industrial purchases, changes in the stock of raw materials, usage of energy and other costs
Value added	Total production value minus total consumption value
Indirect taxes	Sum of indirect taxes (excluding value added tax) and levies less operating subsidies received
Labour costs	The sum of gross wages and salaries paid by the firm
Gross result	Value added minus labour costs and indirect taxes

Although the vast majority (around 89 percent) of the manufacturing firms employs less than twenty workers, the firms in the dataset cover approximately eighty percent of total manufacturing value of production and employment, and about 75 percent of total investments. Hence, the exclusion of firms with less than twenty employees does not have a major effect (in terms of production, employment and investments) on the representation of the Dutch manufacturing sector, at least in a quantitative sense.

Still, the exclusion may affect the typical patterns of entry and exit by small firms, as observed in the empirical literature. For instance, it has often been noticed that hazard rates of entering firms decline with firm age and initial firm size. But these observations are based on empirical studies exploring firm-level databases that include firms of all sizes. Does the exclusion of firms with less than twenty employees affect these and other empirical regularities? The next section will deal with this question.

Table 2.2 Comparison of the SN Database with Total Dutch Manufacturing

	Number of firms		Value of production (mln. guilders)		Employment (×1000)		Investments in fixed assets (mln. guilders)	
Year	Sample	Total manuf.	Sample	Total manuf.	Sample	Total manuf.	Sample	Total manuf.
1978	5444	.	145535	180984	819	998	6595	.
1979	5411	.	160984	201194	815	987	6741	.
1980	5377	.	171630	219283	807	974	7297	.
1981	5209	.	184384	233868	773	949	6691	.
1982	5006	.	185277	240214	733	911	6668	.
1983	4987	44167	192700	248746	705	870	7076	.
1984	4893	43586	212420	274616	695	849	8917	.
1985	4951	44846	224905	280894	709	860	10444	14998
1986	4965	44742	217122	260156	725	872	11959	16328
1987	5080	45735	214838	257907	732	882	12786	16511
1988	5295	46736	229713	272780	732	889	11789	15657
1989	5517	48215	248572	295163	751	902	12955	16070
1990	5593	49240	255370	303753	764	920	13286	17515
1991	5868	50798	258769	309752	765	917	12852	16819
1992	6066	51661	260972	308674	757	910	12725	15856

Note: The numbers of firms for the total manufacturing sector were obtained from the website of SN (http://statline.cbs.nl); before 1983 these numbers were not available. The data on the value of production, employment and investments for total manufacturing were taken from the national accounts statistics, also provided by SN. Before 1985 the investments figures from were not available for the manufacturing sector separately.

2.3 STYLISED FACTS AND THE DUTCH MANUFACTURING SECTOR

As mentioned in the introduction of this chapter, the increased availability of micro-level longitudinal databases has led to a rich set of stylised facts in industrial economics. This section will select a number of these stylised facts and investigate whether they can also be found for the population of Dutch manufacturing firms having more than twenty employees. The selection of stylised facts is based on a number of criteria. The first and most obvious one is that the stylised fact selected should be testable, given the available variables of the SN dataset. Hence, stylised facts involving, for instance, research and development expenditures, patent applications, and advertisement expenditures are excluded in advance. The second criterion is related to the limitations of the SN dataset with regard to the observation threshold. We will try to select and test some stylised facts that should indicate to what extent the regularities emerging from, for instance, the entry and exit process are affected by the exclusion of firms with less than twenty employees. The third and last selection criterion is related to the relevance of the stylised fact for the objectives of this book. Certainly this is an arbitrary task, but given the large amount of empirical research output of especially the last decade a selection had to be made here.

The presentation of the selected stylised facts will be organised as follows. First, we will focus on the regularities observed at the firm level. These include, e.g., survival patterns, persistency in performance levels, and the evolution of firm sizes. Then we will zoom out to study the impact of the firm level regularities at the aggregate level of the manufacturing sector. Here, we will look for stylised facts related to, e.g., size distributions, or the aggregate impact of entry and exit. Finally, we will try to identify the stylised facts at the industry level that have been established so far in industrial economics. Since the aim of the present chapter is to provide a survey of the stylised facts in industrial economics, we will not elaborate on theoretical considerations regarding the observed relationships.

2.3.1 Firm Level

Let us start at the beginning of a firm's life. Obviously, as many studies[4] have shown, the first years are the most difficult for the newborn firm. Entrants are typically small, and commence their operations at relatively low productivity levels (Baldwin, 1995). Thus infant mortality rates are high, but for a given entry cohort exit rates decline over time. Besides the negative relationship between exit rates and age, the initial size of entrants also seems to have a positive impact on the probability to survive, whereas survivors' growth rates are often observed to be negatively related to initial size and age (see Dosi et al., 1997, and Geroski, 1995).

We will first consider the relative size and productivity of entrants. Figure 2.1 displays the ratio of the mean value of the size (number of employees) and productivity (value added per employee) of entrants divided by the mean of all firms in Dutch manufacturing. On average, the size of entrants is 36 percent of the size of existing firms in the first year of birth,

[4]For an extensive overview see Caves (1998).

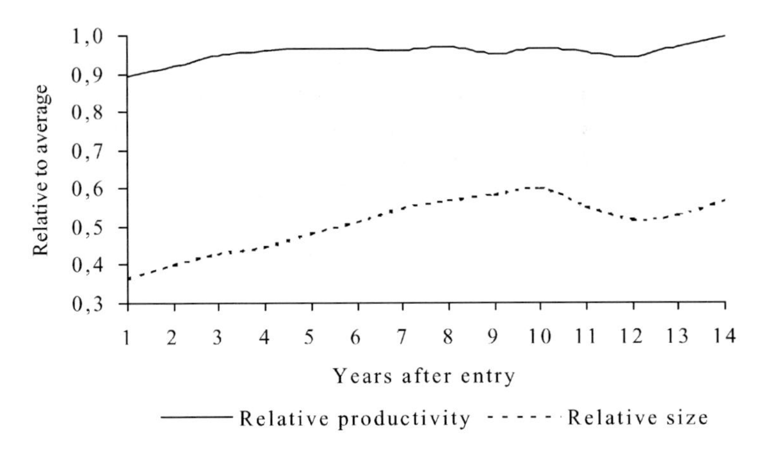

Figure 2.1 Average Relative Size and Productivity of Entrants

gradually growing to 57 percent after 14 years. In a similar exercise, but with a dataset containing firms of all sizes, Baldwin (1995) found entrants to start at 17 percent of the Canadian manufacturing average, increasing to 33 percent after a decade. Further, he found that the productivity of entrants averaged about 73 percent of manufacturing average at birth, increasing to 100 percent after ten years. In our data, we see that the average initial productivity is higher (89 percent of manufacturing average), but after fourteen years entrants are also on average as productive as all manufacturing firms.

Next, we consider the relationship between exit rates and ‘age’. Figure 2.2 shows the average exit rates of entrants versus the number of years after entering the dataset. The pattern is obvious: average exit rates decline as entry cohorts mature. Before investigating the effect of initial size on the probability to survive, let us first have a look on the distribution of the initial firm sizes, presented in figure 2.3. Initial size is measured by the number of employees in the year of first appearance in the dataset.

As figure 2.3 shows, most firms enter the database employing approximately 30 workers. More precisely, the median number of employees at the year of entering is equal to 31, whereas the mean equals 52.5 employees (with a standard error of 1.86). Next, we have estimated the following logit model for all appropriate entry cohorts:

$$Surv = \beta_1 + \beta_2 \ln(ini\_empl) + \varepsilon, \qquad (2.1)$$

where *Surv* = 0 if a firm exits before 1992 and 1 otherwise. The variable *ini_empl* denotes the number of employees of each firm in its year of entry. Table 2.3 shows the regression statistics.

The results of table 2.3 show that there is some mixed evidence for a positive effect of initial size on the probability to survive. For five entry cohorts we have found estimates for β_2 that were not significantly different

Figure 2.2 Average Exit Rates of Entrants by Years After Entry

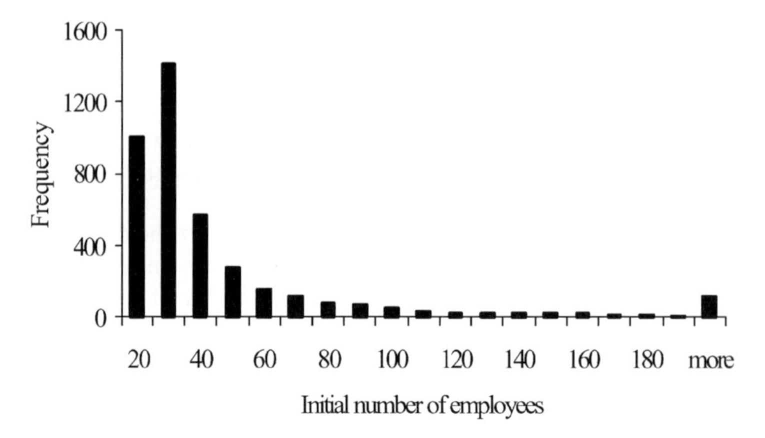

Figure 2.3 Distribution of Initial Firm Sizes of Entrants

Table 2.3 Regression Analysis of Survival on Initial Size

Entry cohort	Constant	Initial size	−2 log L
1979	−1.18 (0.54)	0.23 (0.15)*	614.68
1980	0.51 (0.73)	−0.25 (0.22)	421.16
1981	−0.33 (0.68)	0.08 (0.18)	357.40
1982	−1.32 (0.65)**	0.35 (0.17)**	393.57
1983	−1.06 (0.48)**	0.34 (0.13)***	660.23
1984	−0.73 (0.69)	0.28 (0.18)	404.21
1985	0.26 (0.68)	0.06 (0.18)	386.78
1986	−1.38 (0.86)	0.49 (0.23)**	321.29
1987	−0.62 (0.83)	0.39 (0.23)*	375.68
1988	−0.02 (0.83)	0.25 (0.23)	395.04
1989	−0.96 (0.78)	0.48 (0.22)**	447.95
1990	−2.50 (1.14)**	1.01 (0.33)**	314.01
1991	−1.77 (1.19)	1.07 (0.35)***	398.45
All entry cohorts	−0.13 (0.18)	0.20 (0.05)***	6343.5
Firms present in 1978	−1.63 (0.13)***	0.34 (0.03)***	7364.0

Note: Standard errors in parentheses.
*significant at the 10%-level.
**significant at the 5%-level.
***significant at the 1%-level.

from zero at the 10 percent level.[5] The other entry cohorts did show a significant effect of initial size. When all entry cohorts were pooled together, the effect of initial size was significant at the 1 percent level. In order to compare the regularities of these entry cohorts with all the firms present in 1978, we have run regression (2.1) for these firms as well, where *ini_empl* is the number of employees in 1978. Here, we also found a significant positive effect of size on the probability to survive (see the last row of table 2.3).

Finally, we examine whether initial size and age indeed have a negative impact on the growth rates of surviving entrants. For all surviving entrants (i.e., entrants still present in 1992), we have estimated the following model by ordinary least squares:

$$\ln\left(\frac{empl_t}{empl_{t-1}}\right) = \beta_1 + \beta_2 \ln(ini\_empl) + \beta_3 \ln(age_t) + \varepsilon, \qquad (2.2)$$

[5]This is based on the calculation of the probability of obtaining (by chance alone) a chi-square statistic (for testing the hypothesis that the parameter estimate is zero) greater in absolute value than that observed given that the true parameter is zero.

Table 2.4 Regression Analysis of Post-entry Growth Rates of Surviving Entrants on Initial Size and Age

Constant	Initial size	Age	Adjusted R^2
0.135 (0.010)	−0.024 (0.002)	−0.007 (0.003)	0.010

Note: Standard errors in parentheses.

where $empl_t$ is equal to the number of workers employed by the firm at year $= t$, and age_t is equal to the number of years a firm is present in the dataset at year $= t$. As the results in table 2.4 show, initial size and age indeed have a significant[6] negative effect on the growth of surviving entrants, although the parameter estimates are very small.

In conclusion, we see that the stylised facts related to entering firms can also be found in the SN data on Dutch manufacturing. Despite the uncertainty we have with regard to e.g. the age of firms appearing in the dataset, still the regularities observed in the empirical literature for real entrants are also found for firms entering the SN dataset: exit rates decline as entry cohorts mature, the probability to survive is positively related to the initial size of entrants and finally, the growth of surviving entrants is negatively related to their initial size and age.

For continuing firms, similar studies of patterns of firm growth are found in the empirical literature. Especially those studying the relationship between growth and size are often led by investigations related to the validity of Gibrat's 'Law of Proportionate Effect'. This 'Law' basically states that in each period the expected value of the increment to a firm's size is proportional to the current size of the firm (Sutton, 1997). More formally, let $x(t)$ denote the size of a firm at time t, then Gibrat's Law states:

$$x(t+1) - x(t) = \varepsilon_t x(t), \tag{2.3}$$

where the random variable ε_t denotes the proportionate rate of growth. Most studies[7], however, reject this law. For instance, Evans (1987a, 1987b) found that firm growth and its variance decreases with both firm size and age for a sample of U.S. manufacturing firms. Do these empirically observed departures from Gibrat's Law also apply to Dutch manufacturing firms?

[6]All coefficients are significant at 1 percent level, according to a t-test to test the hypothesis that the parameter is zero.

[7]See Sutton (1997) for an extensive overview.

Obviously, the relationship between the growth of continuing firms (i.e., firms present in 1978 and 1992) and their age cannot be studied here, however we can investigate the relation between growth rates and sizes.

Similar to Evans (1987b), we regressed the *average* annual growth rate of employment for each continuing firm on the logarithm of its number of employees in 1978. Hence, we have

$$[\ln(empl_{1992}) - \ln(empl_{1978})]/14 = \beta 1 + \beta_2 \ln(empl_{1978}) + \varepsilon. \tag{2.4}$$

The ordinary least squares estimate of β_2 gives a value of -0.013, with a t-value of -17.7.[8] Alternatively, we run a similar regression taking the annual growth rates as the dependent variable. Hence, we have

$$\ln(empl_{t+1}) - \ln(empl_t) = \beta_1 + \beta_2 \ln(empl_t) + \varepsilon, \tag{2.5}$$

Here, the estimate for β_2 equals -0.009, with a t-value of -12.6.[9] Hence, both tests corroborate the findings mentioned by Sutton (1997) that the growth of continuing firms declines with their sizes. Consequently, we reject Gibrat's Law, since β_2 is significantly different from zero.[10]

With regard to the relationship between the variance of growth rates and firm size, we calculate the variability of growth rates in the following way. For each firm present in 1978 and 1992, we calculate the standard deviation of the actual annual growth rates (*StdGrowth*). The logarithm of this standard deviation is then used in the following equation:

$$\ln(StdGrowth) = \beta_1 + \beta_2 \ln(empl_{1978}) + \varepsilon. \tag{2.6}$$

Estimating this regression (by ordinary least squares) returns a value of -0.084 for β_2, with a t-value of -8.2.[11] Also here the data corroborate earlier findings: the variance of growth rates of firms significantly decreases with their sizes.

The next stylised fact we have selected concerns the observed heterogeneity of firms. Given the overwhelming amount of evidence regarding firm

---

[8]The parameter estimate for β_1 is equal to 0.058, with a t-value of 18.3; the adjusted R^2 is equal to 0.11.

[9]Here, the parameter estimate for β_1 is equal to 0.042, with a t-value of 12.9; the adjusted R^2 is equal to 0.005.

[10]To see why $\beta_2 = 0$ is consistent with Gibrat's Law, rewrite (2.3) as: $x(t+1)/x(t) = 1 + \varepsilon_t$.

[11]The parameter estimate for β_1 is equal to -1.97, with a t-value of -43.2; the adjusted R^2 is equal to 0.03.

heterogeneity across and within industries, the adjective 'stylised' may very well be omitted here. As Jensen and McGuckin (1997) argue, within industries heterogeneity is pervasive along a wide variety of dimensions, including size, investment patterns, productivity and profitability. We will not show the dispersion of all these variables here, however we will analyse one particular variable (i.e., labour productivity) as we will need it to test the following stylised fact: Not only firms exhibit asymmetric performances, but, equally important, these performance differentials appear to be persistent.[12]

Let us first focus on the dispersion of productivity levels within industries. For each industry, the weighted average of labour productivity is calculated for each year. Then, for each year, each firm's labour productivity is divided by the industry mean to obtain its relative labour productivity (*RelProd*). Figures 2.4a and 2.4b plot the dispersion of these relative labour productivity levels of respectively all firms, and for a sample of continuing firms only. In the computation of the relative productivity of continuing firms, entrants and exiters are also excluded in the calculation of the industry average.

Obviously, firms differ widely with regard to relative productivity levels. Even when the (possibly) low-productive entering and exiting firms are excluded (see figure 2.4b), we still observe a high degree of heterogeneity.

Theoretically, this dispersion could be due to variation of productivity levels over time. For instance, some years a firm may be at 80 percent of the industry average, whereas other years it may perform at 120 percent. However, several studies[13] have shown that these asymmetric performances are persistent. For instance, Jensen and McGuckin (1997), using U.S. census data, show that the productivity distribution is persistent over a five-year interval. We perform a similar regression analysis for Dutch manufacturing firms to investigate the extent of persistence. We regressed the 1992 relative labour productivity of all continuing firms on the similar variable for all preceding years:

$$\ln(RelProd_{1992}) = \beta_1 + \beta_2 \ln(RelProd_t) + \varepsilon. \tag{2.7}$$

Table 2.5 shows the regression statistics. Since all values are positive and significant at the 1 percent level, we can accept the hypothesis that relative productivity differentials are persistent.

---

[12]See Cefis (1998) and Dosi et al. (1997).

[13]See Dosi et al. (1997) for an overview of these studies.

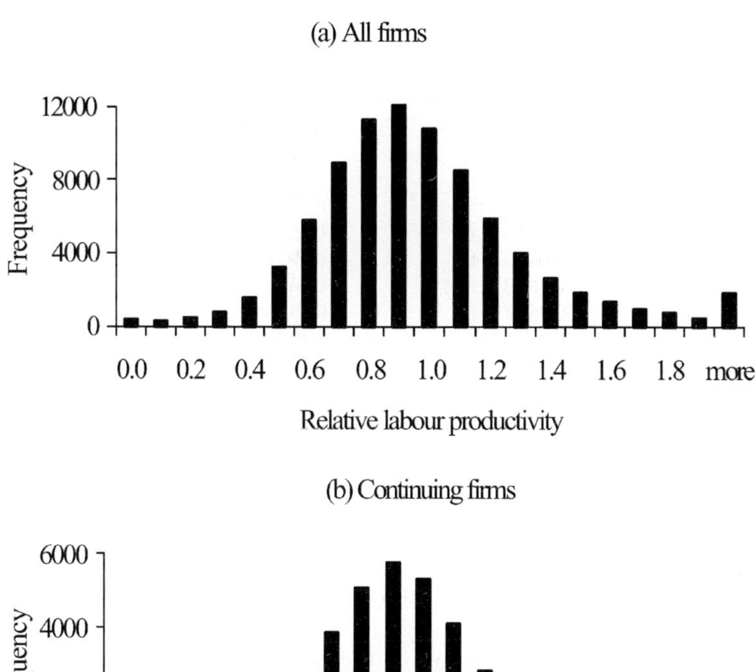

Figure 2.4 Distribution of Relative Productivity Levels

We conclude therefore that the selected stylised facts related to continuing firms are found for Dutch manufacturing firms with more than twenty employees as well. We see that firms' growth rates and the variances of their growth rates fall with their sizes. Further, we found rather strong evidence that the observed differentials with respect to firm performances (measured by labour productivity) are persistent.

Table 2.5 Regression Analysis of Persistence in Relative Productivity Levels

Year (t)	Constant	ln($RelProd_t$)	Adjusted R^2
1978	−0.08 (0.01)	0.38 (0.02)	0.12
1979	−0.08 (0.01)	0.44 (0.02)	0.14
1980	−0.08 (0.01)	0.48 (0.02)	0.17
1981	−0.07 (0.01)	0.52 (0.02)	0.20
1982	−0.07 (0.01)	0.52 (0.02)	0.21
1983	−0.06 (0.01)	0.54 (0.02)	0.22
1984	−0.06 (0.01)	0.56 (0.02)	0.24
1985	−0.06 (0.01)	0.58 (0.02)	0.26
1986	−0.05 (0.01)	0.59 (0.02)	0.29
1987	−0.05 (0.01)	0.63 (0.02)	0.32
1988	−0.05 (0.01)	0.66 (0.02)	0.36
1989	−0.04 (0.01)	0.64 (0.02)	0.39
1990	−0.04 (0.01)	0.76 (0.01)	0.55
1991	−0.02 (0.01)	0.85 (0.01)	0.67

Note: Standard errors in parentheses. All parameter estimates are significant at the 1%-level.

2.3.2 Manufacturing Level

We now turn to an investigation of the regularities in the manufacturing sector as a whole. Again, we start by focussing on the process of entry and exit. Most studies on this subject report substantial entry and exit rates in manufacturing. Cable and Schwalbach (1991) calculated average annual entry rates of about 6.5 percent in terms of number of firms, and about 2.8 percent in terms of market shares for eight countries over various time periods. Average exit rates were found to be roughly similar: 6.5 and 3.0 percent respectively. Table 2.6 shows the entry and exit rates for Dutch manufacturing for 1978 to 1992. Firm entry and exit rates are derived by counting for each year the number of entering and exiting firms and dividing them by the total number of firms active in manufacturing in the same year. Accordingly, the shares of sales (employment) of entering and exiting firms in total manufacturing sales (employment) are calculated for each year in order to measure annual sales (employment) entry and exit rates.

As table 2.6 shows, despite the observation threshold of twenty employees in the SN database, the numbers are strikingly similar to those found in other studies using complete firm level databases. We also find substantial annual entry and exit rates, although the shares of entering and exiting firms in total employment and sales are obviously much smaller, given the relatively small firm sizes of entrants and exiters.

Table 2.6 Entry and Exit Rates in Dutch Manufacturing

	Entry rate (%)			Exit rate (%)		
Year	Firms	Empl.	Sales	Firms	Empl.	Sales
1978	.	.	.	6.67	3.14	2.95
1979	8.39	3.07	2.23	5.41	2.30	1.74
1980	5.82	1.93	1.56	7.14	2.68	1.85
1981	4.95	1.88	1.61	7.85	3.26	4.15
1982	5.73	2.25	1.87	8.83	4.63	3.81
1983	9.73	4.91	3.88	7.20	3.03	2.15
1984	6.09	2.39	2.05	4.93	2.20	1.42
1985	5.88	2.34	1.88	4.34	1.84	1.66
1986	4.89	1.72	1.68	3.99	1.62	1.17
1987	6.06	2.13	1.53	4.09	2.02	1.82
1988	6.14	2.05	1.61	4.15	1.98	1.55
1989	6.54	2.32	1.61	4.91	2.32	1.86
1990	5.17	2.41	1.73	6.74	3.42	2.71
1991	9.36	3.47	2.43	7.06	3.25	2.41
1992	7.27	2.96	2.13	.	.	.
Mean	6.52	2.56	1.98	5.95	2.69	2.23

Given the high hazard rates of entrants, the long run or cumulative impact of entry may be far less substantial. However, as Baldwin (1995) shows for the Canadian manufacturing sector, the accumulation of entry (over 1970–1981) is of a considerable magnitude. The number of entrants alive in 1981 equalled 35.5 percent of the 1970 firm population, while their number of employees was equal to 10.9 percent of total 1970 employment. The magnitude of exit was found to be substantial as well: 35.0 percent in terms of firm numbers, 10.5 in terms of employment.

Calculating these numbers for Dutch manufacturing over 1978–1992 also shows a significant cumulative impact of entry and exit. In 1992, 43.0 percent of all firms are firms that were not present in 1978, collectively capturing 22.4 percent of total 1992 employment and 19.5 percent of total 1992 sales. Hence, although entrants are typically small and have high hazard rates, those that survive are collectively able to obtain a significant market share in time. Further, the number of firms that were present in 1978, but absent in 1992 equals 39.1 percent of the total population of firms in 1978. Their shares in 1978 employment and sales were 23.5 and 23.3 percent, respectively.

To summarise, when measured over the full period (1978–1992) the Dutch manufacturing sector is characterised by high turnover of market shares due to firms entering and exiting the database. However, this is not the only source of turbulence, of course. Within the population of continuing firms, the turnover due to changing market shares may also substantially contribute to the reallocation of market shares. To measure this source of turbulence, we first calculated for each continuing firm the absolute change (between 1992 and 1978) of its share in total sales of the industry in which it is allocated. Next, for each industry we calculated the sum of these changes, and divided this by two. Finally, we computed the average of this number over the 106 industries. Based on these calculations we find that in the Dutch manufacturing sector, on average 15.5 percent of industrial market share is reallocated between continuing firms over the period 1978–1992. This number is very close to the 16.0 percent found by Baldwin (1995) for the Canadian manufacturing sector, measured likewise over the period 1970–1979.

Potentially, the observed turbulence could substantially contribute to aggregate productivity growth at the industry level. If market shares are generally transferred from the less productive to the more productive firms in an industry, aggregate productivity will rise due to this simple Darwinian selection mechanism. However, individual firms may also contribute to aggregate productivity growth by increasing their productive efficiency. As Baldwin (1995) shows, the contribution of turnover to productivity in Canadian manufacturing is substantial when the cumulated or long run impact of entry and exit is taken into account. Depending on the assumptions related to the replacement patterns between entrants, exiters, and incumbents, he found that between 40 to 50 percent of aggregate productivity growth could be attributed to the turnover process. Another study by Haltiwanger (1997)[14] found similar values for all U.S. manufacturing industries, although his decomposition method was slightly different from Baldwin's. In what follows we will decompose aggregate productivity growth in order to calculate the contribution of turnover to aggregate productivity growth in Dutch manufacturing.

Labour productivity at the firm level ($Prod_{i,t}$) is again measured as the real value added per employee. The average labour productivity in 1978 and 1992 is then the output (sales) share weighted sum of the productivity levels of all firms in Dutch manufacturing (denoted by $AvgProd_{1978}$ and $AvgProd_{1992}$). Let $Sh_{cont,t}$ denote the output share (in total manufacturing) of a continuing firm at t, $Sh_{ext,1978}$ the output share of an exiting firm in 1978,

[14]This study decomposed total factor productivity growth, whereas Baldwin (1995) decomposed aggregate growth of labour productivity.

and $Sh_{ent,1992}$ the output share of an entering firm in 1992. Based on Haltiwanger (1997) we decompose growth of average labour productivity ($AvgProd_{1992} - AvgProd_{1978}$) into the following five components:

1. A within effect: within firm productivity growth weighted by initial output shares

$$\sum_{cont} \left[Sh_{cont,1978} \left(Prod_{cont,1992} - Prod_{cont,1978} \right) \right]$$

2. A between firm effect: changing output shares weighted by the deviation of initial firm productivity and initial average manufacturing productivity

$$\sum_{cont} \left[\left(Sh_{cont,1992} - Sh_{cont,1978} \right) \left(Prod_{cont,1978} - AvgProd_{1978} \right) \right]$$

3. A covariance term: the sum of firm productivity growth times firm share change

$$\sum_{cont} \left[\left(Sh_{cont,1992} - Sh_{cont,1978} \right) \left(Prod_{cont,1992} - Prod_{cont,1978} \right) \right]$$

4. An entry effect: the end year share weighted sum of the difference between productivity of entering firms and initial average manufacturing productivity

$$\sum_{ent} Sh_{ent,1992} \left(Prod_{ent,1992} - AvgProd_{1978} \right)$$

5. An exit effect: the initial share weighted sum of the difference between initial average manufacturing productivity and productivity of exiting firms

$$\sum_{ext} Sh_{ext,1978} \left(AvgProd_{1978} - Prod_{ext,1978} \right)$$

Hence, the sum of these five components equals the growth of average labour productivity ($AvgProd_{1992} - AvgProd_{1978}$). The first three components consider the effects of the population of continuing firms, i.e., firms present in 1978 and 1992, whereas the last two components measure the

Table 2.7 Decomposition of Aggregate Productivity Growth

Share in manufacturing productivity growth of:	Netherlands (1978–1992)	United States (1977–1987)
Within effect	57.3	54.4
Between effect	−8.5	−10.3
Covariance effect	20.2	37.6
Entry effect	25.9	18.4*
Exit effect	5.1	

*This value is based on net entry.

effect of the entry and exit process. Exiting firms are here defined as firms present in 1978, but absent in 1992; entering firms are firms present in 1992, but absent in 1978. Table 2.7 lists the contributions of all components in total productivity growth for the Dutch manufacturing sector. For comparison, the last column lists the results for the U.S. manufacturing, taken from Haltiwanger (1997). All values are expressed in percentages.

Since the period captured by Haltiwanger (1997) is different from our period and given the truncation in the SN dataset, these results are not strictly compatible. Still, two interesting similarities emerge. The first one is the negative value of the between effect. This suggests that among continuing firms the selection process on average does not reallocate market shares from initially less productive firms to more productive ones. The second similarity we found between the Netherlands and the United States and Canada concerns the total contribution of the process of market share reallocation to aggregate productivity growth. Since all but the 'within' component capture changes in market shares, this turnover contribution can be calculated as [100 percent – the 'within' share]. For the Netherlands, this equals 42.7 percent, which is similar to the values between 40 to 50 percent found in the earlier mentioned studies. One important difference between the US and Dutch manufacturing is found for the effects of entry and exit. In Dutch manufacturing, especially the entry of firms with high productivity levels substantially contributed to aggregate productivity growth.[15]

[15]This result would undoubtedly have been different in the absence of the observation threshold of twenty employees in our dataset. Compared to greenfield entrants, firms crossing the observation threshold are probably much more labour productive. In a sense, they have already survived a first selection round in which the wheat (efficient greenfield entrants) was separated from the chaff (inefficient greenfield entrants).

Despite the large amount of turnover of market shares due to entry, exit, and the growth and decline of continuing firms, it has often been observed in earlier empirical work that there is a strong persistence over time of a skewed distribution of firm sizes in manufacturing as a whole, following approximately a Pareto distribution (Dosi et al., 1995, Dosi et al., 1997). To see whether a similar distribution is found in Dutch manufacturing, we ranked for each year all firms according to their size (number of employees) in descending order, and plotted to logarithm of their rank (*ln_rank*) against the logarithm of their number of employees (*ln_empl*). Figure 2.5 shows the 15 size distributions (one for each year between 1978 and 1992).

Visual inspection of this figure shows that these distributions indeed approximate a Pareto distribution. Further, the distributions appear to be stable over time, as they are fairly near to each other. However, stronger evidence should come from running the following regressions:

$$ln\_empl_t = \beta_1 + \beta_2\, ln\_rank_t + \varepsilon, \qquad (2.8)$$

where $\beta_2 = -1$ would be consistent with a Pareto distribution. Table 2.8 shows the regression statistics.

Since all estimates are significantly[16] different from -1, the distribution of firm sizes does not perfectly follow a Pareto distribution. Moreover, we observe that the distribution is not perfectly stable over time: after 1984 the distribution becomes gradually less skewed (as the estimate for β_2 decreases in absolute terms). However, the general pattern is clear. For all years the size distributions are fairly close to a Pareto one. One objection that can be made here is related to the high level of aggregation. In the simulations of the model by Dosi et al. (1995), the departures from the Pareto distribution are much larger when individual microsectors are considered. However, random aggregations (of their simulation results) over different microsectors with different size distributions result in distributions much closer to the Pareto one.

Therefore, disaggregating the manufacturing sector into the 106 industries in the SN database may show size distributions less stable and less close to the Pareto distribution than the one for the manufacturing as a whole. Also the other regularities and statistics shown in this section are derived from the manufacturing sector as a whole, concealing possible differences between the industries. In the next section we will disaggregate the manufacturing sector and focus on the differences between the industries.

[16]Based on *t*-test on the hypothesis that $\beta_2 = -1$.

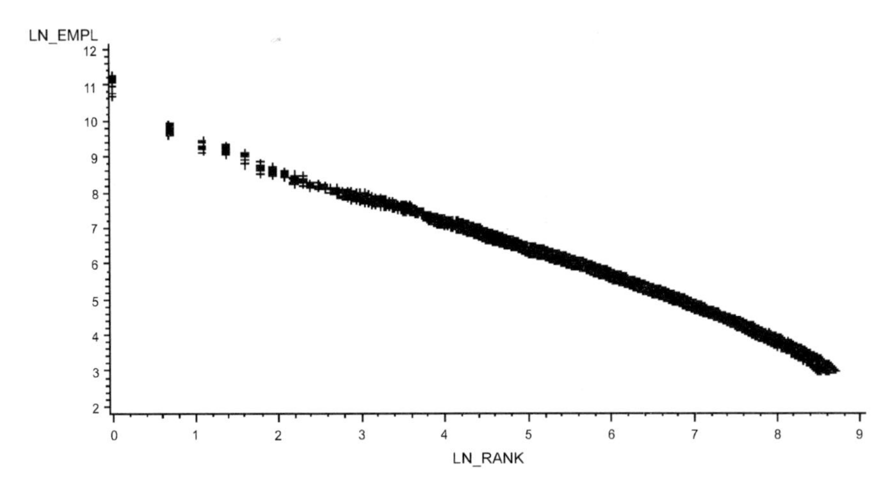

Figure 2.5 Firm Size Distributions (1978–1992)

2.3.3 Industry Level

Given the objectives of this book, the most important stylised fact discussed in this chapter is related to the observed differences between industries. Many studies have observed that variables like capital intensity, advertising intensity, R&D intensity, concentration, and entry and exit rates differ widely across sectors, observations that are indeed at the very origin of the birth of industrial economics as a discipline (Dosi et al., 1997). At the end of this section we will show the specific distributions of a number of selected variables over the industries in the SN database. But first we will investigate whether the data on Dutch manufacturing support cross-sectional relationships that have been observed in earlier empirical research.

Compared to the evidence obtained at the firm and manufacturing level, empirical evidence on cross-sectional relationships at the industry level is rather thin. Most of this evidence addresses the (theoretical) determinants of entry and market share turnover among incumbents, the relation between entry and exit rates, and the determinants of industry-level profitability. As we will show, a considerable part of the evidence is 'negative', i.e., it shows that some explanatory variables have repeatedly been found to be insignificant. Especially in cases where, from a standard theoretical point of view, these variables are supposed to have considerable explanatory powers, negative results are naturally equally interesting.

Table 2.8 Regression Analysis of Size Distributions

Year (t)	Constant	ln_rank_t	Adjusted R^2
1978	11.36 (0.013)	−0.94 (0.002)	0.98
1979	11.36 (0.013)	−0.94 (0.002)	0.98
1980	11.35 (0.012)	−0.94 (0.002)	0.98
1981	11.33 (0.012)	−0.94 (0.002)	0.98
1982	11.27 (0.012)	−0.94 (0.002)	0.99
1983	11.21 (0.012)	−0.94 (0.002)	0.99
1984	11.15 (0.012)	−0.93 (0.002)	0.99
1985	11.12 (0.012)	−0.93 (0.002)	0.99
1986	11.10 (0.012)	−0.92 (0.002)	0.99
1987	11.04 (0.012)	−0.91 (0.002)	0.99
1988	11.01 (0.012)	−0.90 (0.002)	0.99
1989	11.01 (0.012)	−0.90 (0.002)	0.98
1990	11.03 (0.012)	−0.90 (0.002)	0.98
1991	11.02 (0.012)	−0.89 (0.002)	0.98
1992	10.99 (0.011)	−0.89 (0.001)	0.98

Note: Standard errors in parentheses.

But first we will deal with a general phenomenon that has been observed by many studies on entry and exit, i.e., the strong correlation between entry and exit rates (Geroski, 1995). For the Dutch manufacturing sector this relationship exists as well: for the correlation between the industry averages of annual sales entry and exit rates we found a (Pearson) correlation coefficient of 0.58 (0.76 when firm entry and exit rates are correlated). When cumulative entry and exit rates are considered, we also find significantly positive relationships: 0.23 between cumulative firm entry and exit, and a correlation coefficient of 0.52 when cumulative sales entry and exit rates are considered.[17] Hence, at the industry level high entry rates are associated with high exit rates, regardless of how the entry and exit rates are measured.

With regard to the determinants of gross entry, earlier research has shown that profitability does not seem to have a significant effect on attracting

[17]All these correlations have p-values less than 0.01. This p-value (of a correlation r) is obtained by treating the statistic

$$t = \frac{r\sqrt{(n-2)}}{\sqrt{(1-r^2)}}$$

as having a Student's t distribution with (n-2) degrees of freedom, where n is the number of observations. The p-value of the correlation r is the probability of obtaining (by chance alone) a Student's t-statistic greater in absolute value than the observed statistic t.

entrants. Obviously, this observation is at odds with standard models of the entry process, in which high profits are supposed to attract profit-seeking entrants. Mixed results were obtained for the effect of industry growth, whereas capital intensity, scale economies and concentration levels were found to have a negative effect on gross entry.[18] Since we have several ways to measure the entry process, we will analyse the regression results of each of the entry measures on the following 'determinants.' For profitability, we use the industry mean of profits over sales. Since we have no data on capital stock, we will use the industry mean of investments over sales as a proxy for capital intensity. For scale economies, we use a proxy as well, which is (the log of) the median firm size in an industry. Finally, for concentration levels we use the average Herfindahl-index of an industry.[19] Table 2.9 lists the results of the (ordinary least squares) regression analysis.

For all measures of entry we observe a negative effect of average profit margins and median firm size, whereas industrial growth rates always show a positive effect on the entry rates.[20] Investment margins have a significant positive effect on annual firm entry rates, for the other entry measures the effects of investment margins are insignificant. With regard to the Herfindahl-indices, we observe a positive effect on firm entry rates, but a negative effect on sales entry rates (only significant for cumulative entry sales rates). This peculiar result is due to the small number of firms typically found in industries with high Herfindahl-indices. A small number of mostly large firms in an industry (denominator) creates an upward bias of firm entry rates, given an equal number of small entrants (numerator). Therefore, measuring the share of entrants in total industrial sales in general provides more reliable results.

These regularities are not consistent with earlier empirical research. Only the effects of scale economies (approximated by median firm size) and concentration levels (in explaining cumulative sales entry rates) are consistent with earlier findings. However, capital intensity (roughly approximated by investment margins) generally does not have a negative impact on entry rates. And instead of an insignificant effect of profitability on gross entry rates, we observe a significantly negative relationship.

---

[18]See Schmalensee (1989), Malerba and Orsenigo (1994), Dosi et al. (1995), and Geroski (1995).

[19]The Herfindahl-index is calculated as the sum of the squared market shares. Using the collective market share of the four largest firms as an alternative measure for concentration does not significantly change the results.

[20]Obviously, the direction of causality can be discussed here.

Table 2.9 Regression Analyses of Entry Rates and Market Share Turbulence Among Continuing Firms

	Annual entry rates		Cumulative entry rates		Market share turbulence among incumbents
	Firms	Sales	Firms	Sales	
Constant	0.118***	0.119***	1.120***	1.054***	0.027
	(0.024)	(0.024)	(0.122)	(0.125)	(0.054)
Profit margin	−0.421***	−0.299***	−1.881***	−2.030***	0.059
	(0.113)	(0.111)	(0.568)	(0.584)	(0.256)
Industrial growth rate	0.020***	0.011***	0.133***	0.116***	0.002
	(0.004)	(0.004)	(0.019)	(0.019)	(0.009)
Investment margin	0.278*	0.153	0.855	0.657	−0.141
	(0.154)	(0.151)	(0.775)	(0.797)	(0.349)
Median firm size	−0.013**	−0.015***	−0.159***	−0.152***	0.037**
	(0.006)	(0.005)	(0.028)	(0.029)	(0.013)
Herfindahl-index	0.263***	−0.011	0.307***	−0.249*	−0.166**
	(0.022)	(0.022)	(0.112)	(0.115)	(0.051)
Adjusted R_2	0.653	0.210	0.490	0.516	0.075

Note: Standard errors in parentheses.
*significant at the 10%-level.
**significant at the 5%-level.
***significant at the 1%-level.

Of course, an exercise like this should be taken with some caution. Explaining the entry rates derived from the SN database is in fact explaining the extent to which small firms are able to grow and pass the observation threshold of twenty employees (annual entry rates) and the extent to which they are able to survive and gain market share (cumulative entry rates). Therefore, the independent variables mostly reflect the growth opportunities for entrants. Interpreted as such, it can well be explained why we observe a negative relationship between profitability and our 'entry rates'. Perhaps high profits initially attract many entrants, but if many of these entrants have difficulties in surviving they may never exceed the observation threshold of twenty employees. And the more difficult it is for entrants to survive and grow, the less competition the incumbents face and hence the more they are able to retain their high profit levels, ceteris paribus. In this interpretation, high profits simply reflect the appropriability conditions in an industry and, consequently, the opportunities for young and small firms to survive and grow.

With regard to the stylised facts related to the determinants of turbulence among incumbents[21], earlier empirical research[22] has shown that variables related to the level of international competition and conditions related to technology and to the demand side explain a great deal of the differences among industries in incumbents' market share mobility. Since these variables cannot be extracted from the SN dataset, the only appropriate variables we can include in the analysis are those we also included in analysing the entry rates. But the evidence on the explanatory powers of these variables is rather thin. No study so far has convincingly shown that any of these variables significantly explains the amount of turbulence among incumbents in an industry. As shown by the low adjusted R_2, displayed in the last column of table 2.9, in Dutch manufacturing these variables together do not explain much of the variation of incumbents' mobility in market shares as well. Only median firm size (positive) and the Herfindahl-index (negative) are statistically significant in explaining the market share turbulence among incumbents.

Another widely studied cross-sectional relationship concerns the determinants of industry level profitability, usually including concentration levels, scale economies and capital intensity. Based upon a survey of empirical literature, Schmalensee (1989) concludes that the relation between seller concentration and industry-level profitability is usually statistically weak or even absent, whereas scale economies and capital intensity tend to be positively related with industry-level profitability. Again, the 'negative' result found for the effect of concentration is interesting, as it contradicts the textbook models of industrial organisation, where high concentration is supposed to reflect market power which, in turn, should lead to high profits.

For investigating this relationship with the SN database, we have regressed the industry-level profit margin on the Herfindahl-index, the median firm size and the investment margin (the variables are defined in the analysis of the determinants of gross entry rates above). Table 2.10 shows the results. In line with the findings reported by Schmalensee (1989), the concentration levels of the industries do not significantly explain the variances in profit margins, whereas capital intensity (approximated by investment margins) is significantly positively related to industry-level profitability. Finally, scale

[21]Measured as half the sum of the sum of the absolute changes in market shares for all continuing firms in an industry between 1978 and 1992.
[22]See Baldwin and Rafiquzzaman (1995) and Baldwin and Caves (1998).

Table 2.10 Regression Analysis of Profit Margins

Constant	Herfindahl-index	Median firm size	Investment margin	Adjusted R_2
0.017 (0.021)	0.015 (0.019)	0.002 (0.005)	0.986*** (0.094)	0.509

Note: Standard errors in parentheses.
***significant at the 1%-level.

economies (approximated by the log of median firm size) are not statistically significant.

The final observation we will focus upon in this chapter addresses the differences between industries with regard to their structural and dynamic properties. Since the aim of this book is to explain the observed cross-sectional variations, this observation is in fact the starting point of the present book. However, before exploring and testing some selected theoretical contributions concerned with explaining the diversity in sectoral patterns in the following chapters, we consider it useful to first give an impression of the extent to which the industries in the sample diverge.

For this purpose, we have selected a number of key-variables that together capture a wide array of dimensions of industry properties, including measures of concentration levels, size distributions, profit and investment margins, and the entry and exit process. Table 2.11 shows the distribution statistics of the selected variables. These variables can roughly be categorised into two groups: one group measures the static properties of industries, the second focuses on more dynamic characteristics of the industries. The first group contains two concentration measures: the average Herfindahl-index (multiplied by one hundred) and the average collective market shares of the four largest firms (CR-4) in each industry. Further, we included the median firm sizes (in terms of number of employees) of each industry, calculated over 1978–1992. For displaying differences in the size distribution, we have calculated the average parameter estimates for $ln\_rank_{i,t}$, resulting from running the following regression (based on regression (2.8)):

$$ln\_empl_{i,t} = \beta_1 + \beta_2 \; ln\_rank_{i,t} + \varepsilon. \tag{2.9}$$

Hence, we ran this regression for each industry i and for each year t, and calculated for each industry the mean estimate of β_2 over 1978–1992. The final two variables in the group of static measures are the industry profit and investment margins. For each firm we calculated the annual mean of its ratio

Table 2.11 Distribution Statistics on Selected Industry-Level Key Variables

Variable	Mean	Std. Err.	Minimum	Maximum	Skewness	Kurtosis
Static measures:						
Herfindahl-index	13.0	1.29	0.95	98.2	3.28	16.1
CR-4	51.2	2.15	11.6	100	0.25	−0.72
Median size	80.3	5.75	33	390	2.60	8.32
Ln_rank	−1.06	0.04	−3.47	−0.49	−2.48	9.27
Profit margin	9.02	0.32	2.87	22.1	0.86	1.76
Investment margin	6.28	0.23	1.76	14.0	0.95	0.86
Dynamic measures:						
Annual entry sales rate	4.13	0.28	0.16	17.3	1.55	3.66
Annual exit sales rate	4.39	0.34	0	19.6	1.73	4.19
Cumulative entry sales rate	28.7	1.90	0	81.6	0.58	−0.39
Cumulative exit sales rate	29.0	2.09	0	82.1	0.63	−0.38
Survival rate	71.3	1.79	0	1	−1.09	2.90
Market share turbulence	15.5	0.61	0.60	35.7	0.66	1.18
'Within' share	40.4	8.53	−3.71	1.62	−3.20	12.6
Industrial growth rates	42.4	6.80	−0.71	3.33	1.54	4.03

Note: Except for the Herfindahl-index, median size and ln_rank, the (unweighted) means of all variables are expressed in percentages. In calculating the average 'within' share, we have only included industries with positive productivity growth. This restriction ruled out 16 industries.

of profits (i.e., gross result) over sales and the ratio of its investments in fixed assets over sales. Table 2.11 shows the statistics of the industry means (calculated over the period 1978–1992) of these ratios.

The group of dynamic measures includes the following variables. Entry and exit is measured by annual and cumulative sales entry and exit rates. Further, we have included survival rates, measured by the ratio (for each industry) of successful entrants (i.e., firms present in 1992, but absent in 1978) over all firms that enter the dataset after 1978, and the market share turbulence among incumbents. The 'within' share, which measures the contribution of individual productivity increases by incumbents to total industry productivity growth, is also included. It is simply the 'within' effect as described in the previous section (within firm productivity growth weighted by initial output shares in the industry), divided by the total productivity growth in an industry between 1978 and 1992. The final row of table 2.11 displays the distribution statistics of the industrial growth rates: the difference in total real sales between 1978 and 1992 of an industry, divided by its total real sales in 1978.

The first four columns of table 2.11 list the mean, its standard error and the lowest and highest values of the selected industry-level variables. In order to get a first impression of the shape of the distribution of these variables over the industries, column three and four show respectively the skewness and the Kurtosis-level for each variable.[23] In Appendix 2 of this chapter the histograms visualise the distribution of the selected variables.

As table 2.11 shows, the various measures related to the static and dynamic characteristics of industries all show a substantial degree of dispersion.[24] Certainly, some variables exhibit distributions that are more symmetric and with heavier tales than others, but all variables display enough variation over industries to conclude that industries differ in all dimensions, whether they are related to more static measures or to more dynamic measures. In conclusion, this last stylised fact tested in this chapter emerges from the SN database on Dutch manufacturing as well.

Although we have not obtained results consistent with earlier findings on the relationship between entry rates and a number of other variables, this section has shown that other regularities at the industry level in Dutch manufacturing are fairly consistent with the stylised facts established so far. We have observed rather strong correlations between the average entry and exit rates of industries. Further, we have found that market share turbulence

[23]For measuring the skewness of the distribution of a variable y with standard deviation S_y and sample size n we calculate:

$$\frac{n}{(n-1)(n-2)} \cdot \sum_{i=1}^{n} z_i^3, \text{ where } z_i = \frac{y_i - \bar{y}}{S_y},$$

which is a measure of the tendency of the deviations from the mean to be larger in one direction than in the other. A positive value for skewness indicates the data are skewed to the right. A negative value indicates the data are skewed to the left. A zero value indicates the distribution is symmetric. Skewness can thus be interpreted as a tendency for one tail of the population to be heavier than the other. Kurtosis is a measure of the heaviness of the tails of a distribution, calculated as:

$$\frac{n(n+1)}{(n-1)(n-2)(n-3)} \cdot \sum_{i=1}^{n} z_i^4 - \frac{3(n-1)^2}{(n-2)(n-3)}.$$

Large values of Kurtosis indicate the distribution has heavy tails.

[24]The strong outliers observed for the 'within' variable are inherent to the applied decomposition method in cases where some industries exhibit low levels of aggregate productivity growth. For instance, the appearance of a large entering firm with a low productivity level may supress the aggregate productivity growth such that a moderate productivity change by incumbents becomes extremely high when it is expressed as a share in aggregate productivity growth.

among incumbents is hard to explain with the standard industry-level variables. This finding is also consistent with earlier empirical research, which highlighted underlying conditions of technology and demand as the important determinants of incumbents' mobility rather then variables like concentration and capital intensity. Next, and again in line with earlier empirical research, we have found that concentration does not significantly explain industry-level profitability. Capital intensity (approximated by investment margins) explains much stronger the variance in the latter variable. Finally, we have observed a high degree of heterogeneity in the properties of the industries in the sample. Both static as well as dynamic measures of the industry characteristics show a wide variety, in most cases displaying rather skewed distributions over the industries.

2.4 CONCLUSIONS

Given the empirical nature of this book and the extensive use of the SN manufacturing database throughout it, this chapter illuminated three important issues related to empirical research in industrial economics. First of all, we have described in detail the firm-level data on which a major part of the research in this book is based. Second, by presenting a number of selected stylised facts we aimed to highlight some robust empirical results that have been established so far in industrial economics. Especially in the last fifteen years, the increased access to firm-level data has resulted in a collection of results that were found to be rather invariant across countries and over time. Finally, we have investigated whether these stylised facts could also be found in Dutch manufacturing.

Two important results emerge from this chapter. The first is that the regularities emerging from the SN manufacturing database are to a large extent consistent with the stylised facts in industrial economics. Whether we look at firm-level regularities, their aggregate impact on the manufacturing sector, or regularities at the industry level, practically all of them seem to be consistent with earlier empirical findings in industrial economics. The second result is in fact already captured by the first one, but is still mentioned separately here because of its essence for the research in this book. We have shown that industries widely differ in virtually all dimensions observable in the data. Certainly, this result is far from new. In fact, it constitutes the origin of industrial economics as a discipline. However, despite the presence of a number of theoretical contributions that potentially can explain the observed differences, not much effort has been done so far in testing these theories with the increasingly available firm-level data.

As mentioned already in chapter one, the aim of this book is to fill this gap. In chapter 4 we will use the SN database to empirically test some theories that may explain the observed differences between industries with regard to their structural and dynamic properties. But first, the next chapter will elaborate on these theoretical contributions.

APPENDIX 1: LIST OF INDUSTRIES IN THE SN DATABASE

Food Products

Production, processing and preserving of meat and meat products
Manufacture of dairy products
Processing and preserving of fish and fish products
Manufacture of grain mill products
Manufacture of vegetable and animal oils and fats
Processing and preserving of fruit and vegetables
Manufacture of bakery products
Manufacture of sugar, cacao, chocolate and sugar confectionery

Beverages, Tobacco, and Food Nec*

Manufacture of prepared animal feeds
Manufacture of other food products nec
Distilling, rectifying and blending of spirits; ethyl alcohol production
Manufacture of malt liquors and malt
Manufacture of soft drinks
Manufacture of tobacco products

Textile Products and Wearing Apparel

Preparation and spinning of wool fibres, weaving of wool
Preparation and spinning of cotton fibres, weaving of cotton
Manufacture of tricot and stockings

Finishing of textiles
Manufacture of carpets and rugs
Manufacture of made-up textile articles, except apparel
Manufacture of other textiles nec
Manufacture of wearing apparel, dressing and dyeing of fur

Leather Products

Tanning and dressing of leather
Manufacture of luggage, handbags and the like, saddlery and harness
Manufacture of footwear

Wood and Wood Products

Sawmilling and planing of wood, manufacture of veneer sheets, plywood, laminboard, particleboard and other panels and boards
Carpentry and manufacture of densified wood and parquet flooring blocks
Manufacture of wooden containers
Manufacture of other products of wood, manufacture of articles of cork, straw and plaiting materials
Manufacture of furniture, except metal furniture

Paper and Paper Products

Manufacture of paper and paperboard
Manufacture of other articles of paper and paperboard
Manufacture of corrugated paper and paperboard

Printing and Publishing

Printing of newspapers
Printing of books
Offset printing
Chemigrafical and fotolithografical firms
Other printing
Publishing of newspapers
Publishing of periodicals

Publishing of books
Other publishing
Bookbinding

Chemicals, Chemical Products, Yarns and Fibres

Manufacture of fertilisers
Manufacture of synthetic resin
Manufacture of dyestuffs and colouring matters
Manufacture of chemical raw materials nec
Manufacture of paints, varnishes and similar coatings, printing ink
Manufacture of pharmaceuticals and medicinal chemicals
Manufacture of soap and detergents, cleaning and polishing preparations, perfumes and toilet preparations
Manufacture of chemical pesticides
Manufacture of other chemical products nec

Rubber and Plastic Products

Manufacture of rubber products
Manufacture of plastic products

Building Materials, Earthenware and Glass

Manufacture of bricks and tiles
Manufacture of ceramics
Manufacture of cement and lime
Manufacture of articles of concrete and cement
Manufacture of other non-metallic mineral products nec
Manufacture of glass and glass products

Basic and Fabricated Metal Products

Manufacture of basic metals
Iron, steel and non-ferrous metal foundries
Forging, pressing, stamping and roll-forming of metal

Manufacture of metal fasteners, cables, springs and the like
Manufacture of tanks, reservoirs and pipelines
Manufacture of steel and non-ferrous metal doors, windows, walls and the like
Metal construction nec
Manufacture of metal furniture
Manufacture of metal packings
Manufacture of heating and boilers, except electrical
Manufacture of other fabricated metal products nec
Forging, treatment and coating of metals

Machinery

Manufacture of agricultural machinery
Manufacture of tools and machinery for metallurgy
Manufacture of machine tools
Manufacture of machinery for packing and wrapping
Manufacture of machinery for food, beverage and tobacco processing
Manufacture of machinery for petrochemical, chemical and pharmaceutical industries
Manufacture of machinery for manufacturers of rubber and plastic products
Manufacture of lifting and handling equipment
Manufacture of machinery for mining, construction, building materials and metallurgy
Manufacture of bearings, gears, gearing and driving elements
Manufacture of machinery for wood and furniture
Manufacture of machinery for textile and apparel
Manufacture of machinery for chemical cleaning, washing, leather and leather products, paper and paper products and printing
Manufacture of engines and turbines, except aircraft, vehicle and cycle engines
Manufacture of office machinery
Manufacture of pumps, compressors, taps and valves
Manufacture of fans, refrigerating and freezing equipment
Manufacture of weighing machinery and domestic appliances, except electrical
Appendage

Manufacture of machine parts nec
Manufacture of machinery nec
Machine repair nec

Electrical Products

Manufacture of insulated wire and cable
Manufacture of electric motors, generators and transformers
Manufacture of electricity distribution and control apparatus
Manufacture of other electrical equipment nec

Transport Equipment

Manufacture of motor vehicles
Manufacture of trailers and semi-trailers
Manufacture of bodies for motor vehicles
Manufacture of parts and accessories for motor vehicles
Building and repairing of ships and boats
Manufacture of motorcycles and bicycles
Manufacture and repair of aircraft
Manufacture of transport equipment nec
*Not elsewhere classified

APPENDIX 2: HISTOGRAMS OF THE INDUSTRY-LEVEL VARIABLES SELECTED IN SECTION 2.3.3

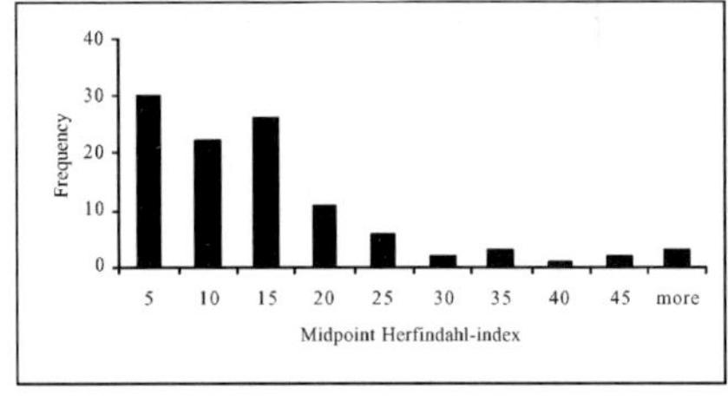

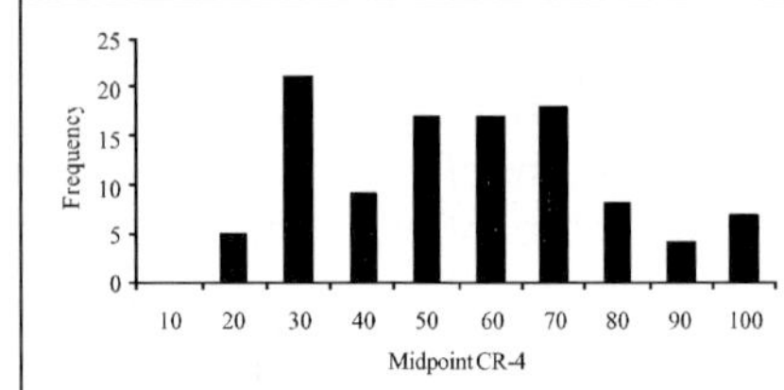

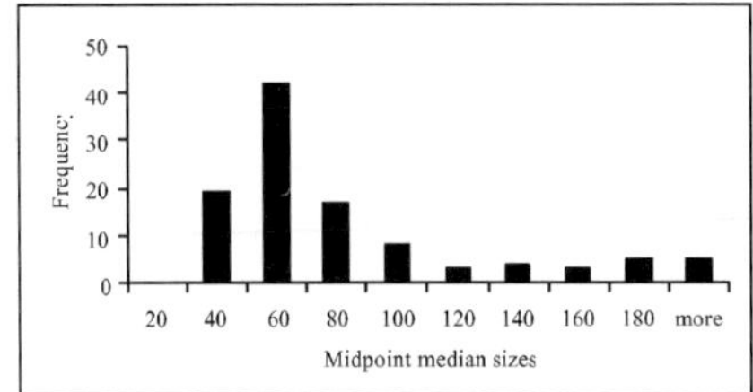

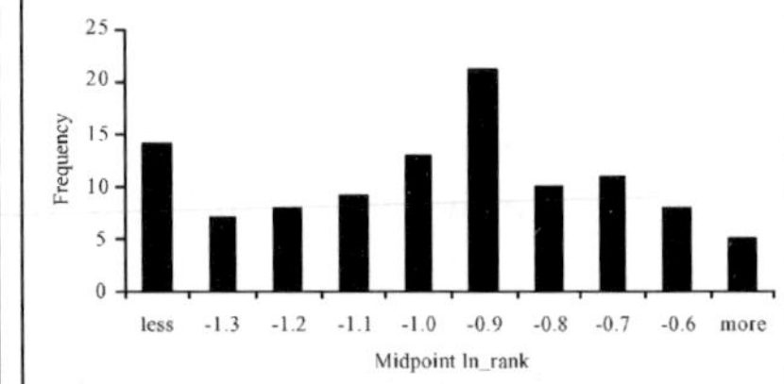

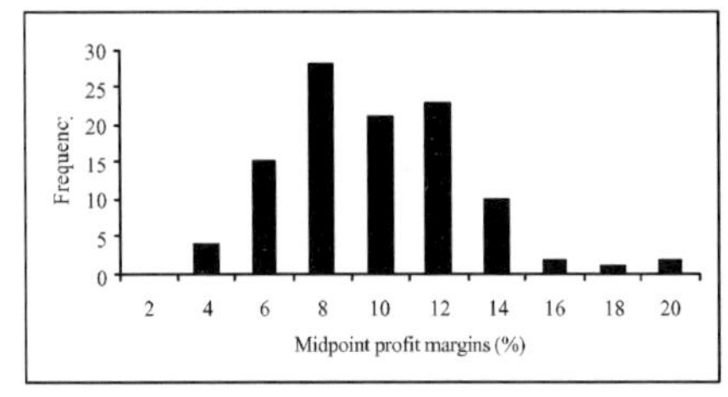

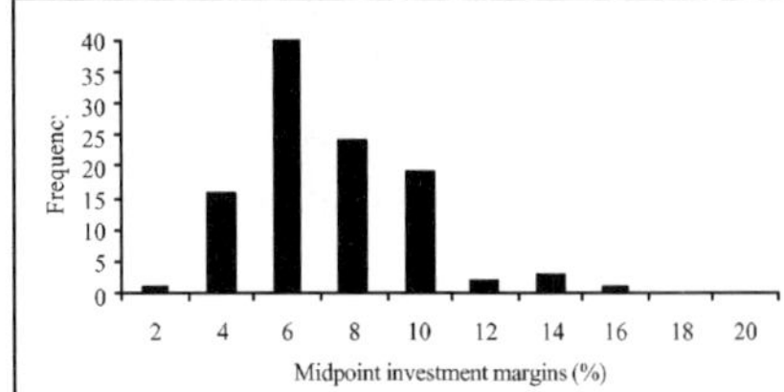

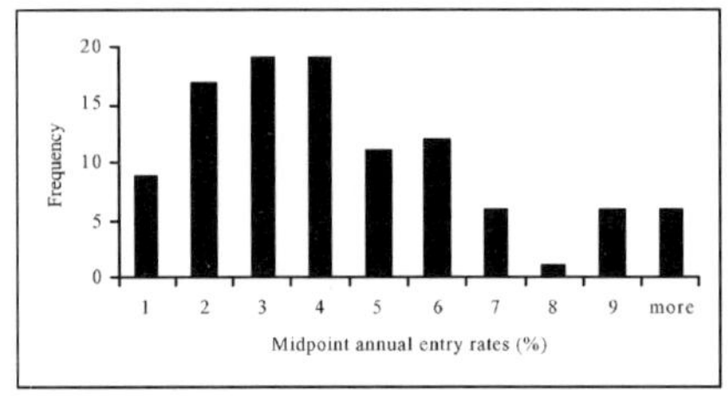

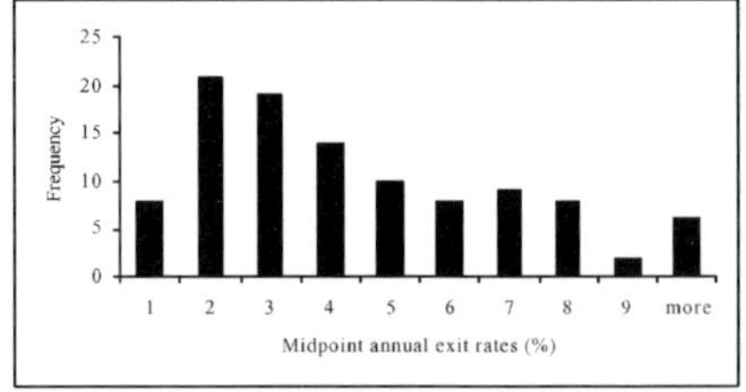

(*Continued*)

(*Appendix Continued*)

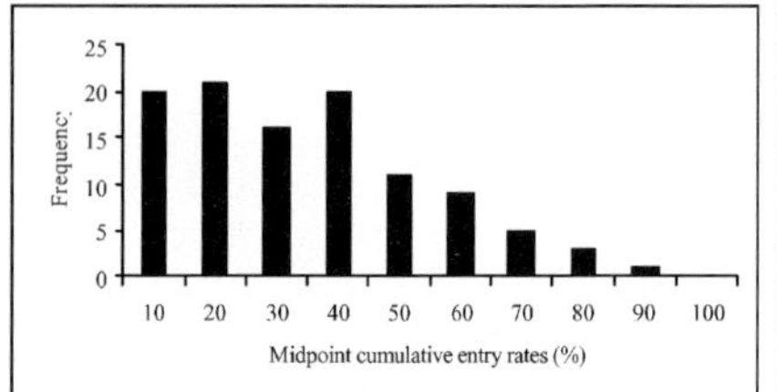

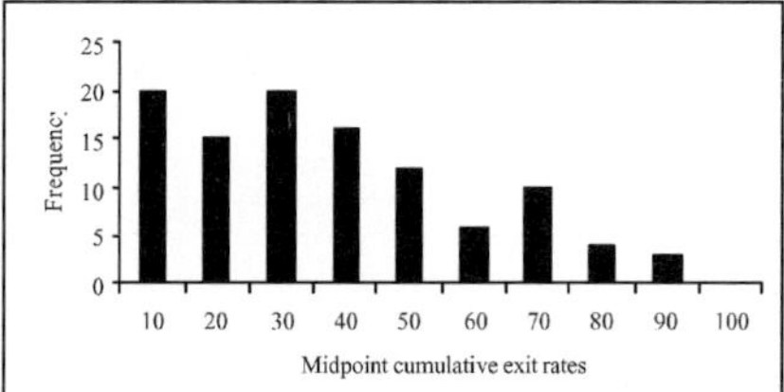

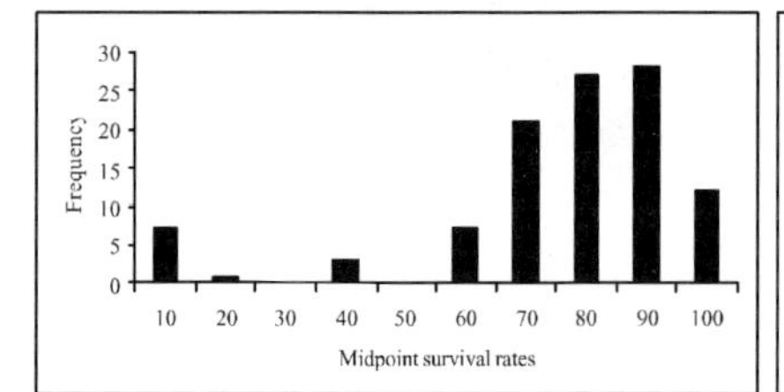

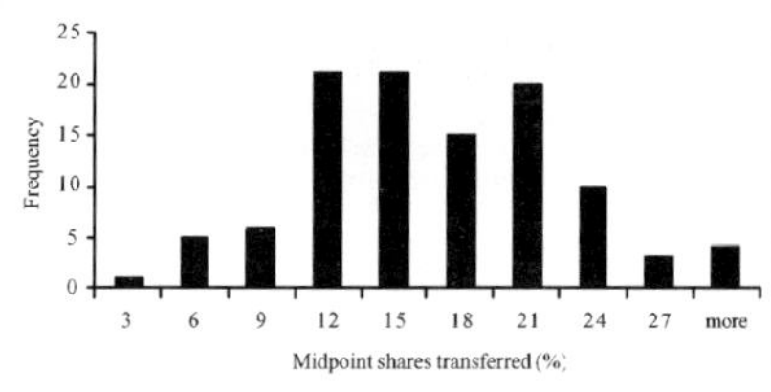

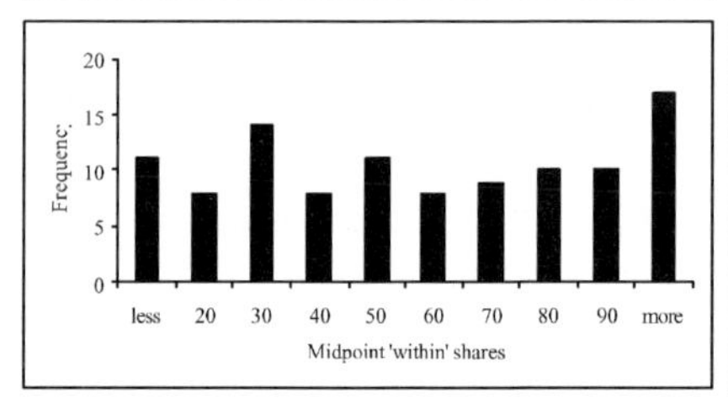

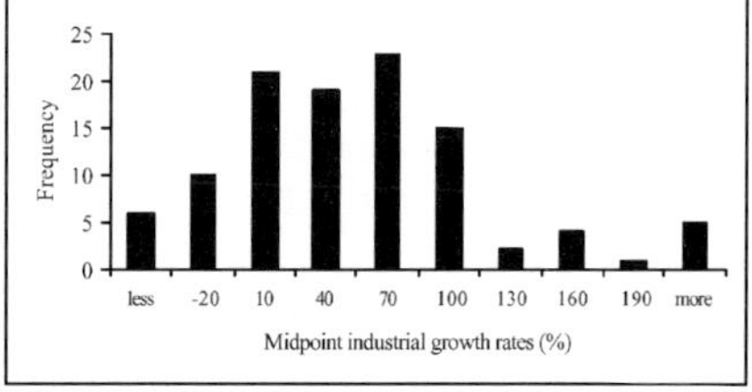

3. SURVEY OF SELECTED THEORIES

3.1 INTRODUCTION

In the early days of industrial economics as a discipline, competition within industries was clearly viewed as a dynamic process. Although Alfred Marshall has laid the foundations of the partial equilibrium analysis found in many textbooks, he acknowledged that dynamics, rather than statics, should be the main concern of economics. Marshall's view is best reflected by his famous analogy of the young trees in the forest of which many succumb, but those few that survive gradually grow and replace the older ones. And as in the forest, "*... in almost every trade there is a constant rise and fall of large businesses ...*" (Marshall, 1936, p. 316).

By describing competition as a process mainly involving the dynamic innovations of the entrepreneur, Joseph Schumpeter (1912, 1942) even more rejected the idea of competition as a static concept. In Schumpeter's view, economic processes are organic: changes come from within the system. In this system, the entrepreneur is a key figure, an active agent of economic progress who induces economic development by actively searching for and carrying out 'new combinations' in production (Ekelund and Hébert, 1990).

Marshall and Schumpeter's acknowledgements of the dynamics of the competitive process were virtually discarded in the subsequent developments in industrial economics. Instead, priority was given to analytical tractability and, consequently, attention shifted to essentially static equilibrium analyses of industrial structures, which constituted the dominant paradigm in industrial economics for many decades. However, the research methodology underlying this paradigm inherently involves some theoretical and empirical limitations. Especially technological change, generally acknowledged to be a major engine driving economic growth, is difficult to embody properly in equilibrium models. In section 2 of this chapter we will focus on these models and explain how the equilibrium approach limits the theoretical and empirical analysis of industrial economics.

Perhaps because of these limitations we recently observe a renewed interest in disequilibrium phenomena that is starting to change the conventional way of analysing industries (Dosi et al., 1997). The recently obtained empirical evidence on the persistence of heterogeneity and turbulence, as discussed in the previous chapter, undoubtedly contributed to this renewed and increasing interest. The remainder of this chapter will focus on two of these rather recently developed theoretical approaches that may help to explain the observed differences between industries with regard to their structural and dynamic properties. The first one, which will be described in section 3, is the framework of technological regimes. A technological regime can be defined as a particular combination of opportunity, appropriability, cumulativeness conditions and properties of the knowledge base underlying the innovative and productive activities of an industry. Given differences in these conditions, different patterns of innovative activities are likely to emerge which, in turn, may affect the structures and dynamics of industries. However, cross-sectional differences between industries may also stem from the different evolutionary stages these industries occupy. Explanations of this type can be derived from theories on product or industry life cycles. This is the second recently developed theoretical approach upon which we will focus in this chapter.

3.2 EQUILIBRIUM MODELS

This section aims to give a brief overview of the equilibrium approaches in industrial economics. It is divided in four parts. First we will describe the static equilibrium models, including the structure-conduct-performance approach and the game-theoretical contributions. Second, we will pay attention to the 'bounds' approach, developed by John Sutton. Because of its explicit aim to construct an empirically testable framework, applicable to a wide range of industries, we have devoted a complete subsection to this approach. Third, we will focus on the class of dynamic equilibrium models. Fourth, and finally, we will discuss the theoretical and empirical limitations of the equilibrium models in industrial economics.

3.2.1 Static Equilibrium Models

For a number of decades the structure-conduct-performance (SCP) approach, developed by Mason and Bain[25], was the dominant paradigm in industrial economics. Based upon neo-classical theory, as well as upon empirical work[26], this paradigm postulates causal relationships between the structure of an industry, the conduct of firms in the industry, and their economic performance. Structure refers to the environment within which firms operate, and includes elements like the number and size distribution of buyers and sellers, the level of product differentiation, and the height of entry barriers. The fact that structural conditions may provide the latitude to act in some other way than being a 'price-taker' in the market leads to the assumption that structure influences conduct. The variety of behaviours that might be realised due to weak structural foundations for competition might, in fact, be very large and include collusive behaviour, price discrimination, et cetera. Regardless of the non-competitive behaviour employed, the assumption of the SCP approach was that, because firms are profit maximising, the result would be detectable differences in the performance of these firms in the market and, in particular, in the allocative inefficiency compared to perfect competition.

Hence, in its simplest form, the exogenously determined structural properties of an industry govern the conduct of firms, which in turn determines the economic performance of a market. More sophisticated versions of this framework also recognise causal relationships in opposite directions. For instance, entry-deterring practices by existing firms (conduct) are aimed to discourage entrants and will thus affect the ultimate structure of an industry. Also, the level of product differentiation is likely to be affected by the marketing strategies of the firms. Although recognising the interaction between the elements has made the SCP paradigm more realistic, it has also posed new problems related to the empirical work in this area. As Schmalensee (1989) argues: "*... in the long-run equilibria with which cross-section studies must be primarily concerned, essentially all variables that have been employed in such studies are logically endogenous. This means that there are in general no theoretically exogenous variables that can be used as instruments to identify and estimate any structural equation.*" (Schmalensee, 1989, p. 954).[27]

[25]See Bain (1956) and Mason (1957).
[26]For instance, see Bain (1951, 1954), or Weiss (1974) for a survey.
[27]The empirical foundations of the structure-performance link were also attacked by a variety of studies contesting Bain's original findings. For an overview, see chapter 9 of Scherer (1980).

Mostly due to this fundamental endogeneity problem, the interests of the mainstream industrial economists shifted from empirical research to more theoretical founded contributions. Most of these theoretical contributions came from the Chicago School during the 1970s and from the New Industrial Economists during the 1980s. Whereas the starting point of the SCP paradigm was imperfect competition, the Chicago School asserts that perfect competition has substantial explanatory power (Martin, 1994). In their view, market power may only exist temporarily, unless supported by government interference. A particularly important point of contention of the Chicago School was the presumed effect of a competitive 'fringe', even in industries that are structurally highly concentrated. Bork (1978), for example, criticised many of the classic antitrust decisions because of their failure to consider the impact of competitors, even though they held a relatively small share of the market. In essence, the argument of the Chicago School was that structure reflected relative efficiency and that larger firms were large because they were more efficient than rivals.[28] If, however, these larger firms proceeded to the 'conduct' stage of the SCP model in ways that prove antithetical to competitive behaviour, the smaller firms would have a significant impact on the market.[29] By breaking the link between structure and conduct, the Chicago school undermined the theoretical consistency of the SCP model.

Subsequent research in industrial economics in the 1980s shifted back the attention from perfect markets to essentially oligopolistic market structures. This school, labelled New Industrial Economics or New Industrial Organisation, is even more based on neo-classical theory than the SCP tradition. However, New Industrial Economists consider conduct as the key element rather than structure. This emphasis on behaviour is reflected by the extensive use of game theory[30] in New Industrial Economics. The strength of these game-theoretical models is the ability to formally incorporate in an elegant way the strategic interactions between players (usually firms in an oligopolistic setting). For instance, duopoly models, such as the Cournot and Stackelberg (leader-follower) models, are set up as non-cooperative

[28]See Demsetz (1973).

[29]A similar argument is derived from the theory of contestable markets (Baumol et al., 1982). Assuming that there are no sunk costs and that an existing dominant firm will not decrease its price if an entrants comes in, this theory claims that the only way the dominant firm can keep an entrant out is to set a price that yields zero economic profit. Hence, even in highly concentrated markets the force of potential competition is enough to yield the same performance as a competitive market (Martin, 1994).

[30]Initially already introduced in economics in the 1940s by Von Neumann and Morgenstern (1944).

games in which the players' strategies consist of setting quantities as a function of the other player's quantity setting. The outcome of such a game (in terms of, e.g., market shares) is a Nash-equilibrium, i.e., a configuration of strategies in which no player has an incentive to change its strategy given the other players' strategies. Other applications of game theory in industrial economics include models of collusive behaviour, product differentiation, innovation and patent races, and predatory pricing strategies.[31]

Despite the novelty of formally embodying strategic interactions, these applications of game theory have not enabled us to say much more on the working of oligopolistic markets than that could be said on the basis of earlier theory (Fisher, 1989). Many outcomes are still possible. Another objection is that the results of such game-theoretic analyses tend to depend delicately on the precise form of the underlying game (Sutton, 1991). Further, these form-specific factors, such as the information partition among the players, are usually difficult to map into empirical categories that hold across a range of different industries.

One way to overcome this latter difficulty is to tailor-make a model to fit a more narrowly defined environment. Such a modelling strategy is likely to increase the degree of precision of predictions, but by definition it will limit the breadth of applications. Acknowledging this trade-off, Sutton (1991, 1998) aims to find a middle way that has come to be known as the 'bounds' approach. Rather than predicting some unique equilibrium outcome, this approach aims to predict a set of outcomes that can be supported as an equilibrium outcome of a wide class of models that may all be plausible a priori. In game-theoretical terms, such an exercise "*...results in a partitioning of the space of outcomes into two sets: the set of points that can be supported as equilibria...and the set of points that cannot.*" (Sutton, 1998, p. 7). For instance, with regard to market concentration, this approach aims to specify a lower bound to the equilibrium level of concentration that holds across a broad set of different industries instead of predicting precise levels of market concentration for a limited set of industries.

Because the 'bounds' approach strongly aims to construct an empirically testable framework, applicable to a wide range of industries, this approach potentially fits well in our search for theories that contribute in explaining the observed differences between industries. Furthermore, this approach is supported by substantial empirical evidence. We consider it appropriate therefore to explain this approach in more detail.

---

[31]Chapter 13 of Rasmussen (1989) gives on overview of these applications.

3.2.2 The 'Bounds' Approach

The central focus of the theory presented in Sutton (1991) lies in disentangling the way in which exogenous and endogenous elements of sunk cost interact with each other in determining the equilibrium pattern of industrial structure (mainly industrial concentration). Industry equilibrium is modelled in terms of a two-stage game, in which the fixed costs of the first stage are treated as sunk costs in the second stage of the game. Sutton distinguishes three cases. The first two cases have in common that the only sunk costs involved are the exogenously given set-up costs. The distinction between them relates to the heterogeneity of the goods produced. If the goods are homogeneous, Sutton shows that as the size of the market increases, the equilibrium number of entrants increases, leading to a continuous decline of concentration levels. This is due to the following process. Given that entry occurs up to the point at which the profits (at stage 2) of the last entrant cover its sunk costs incurred at stage 1, any increase in the market size (starting from any given concentration level) will raise profits and thus attract more entrants.

Multiple equilibria arise in case of heterogeneous goods, with sunk costs still being exogenous. Sutton argues that under these circumstances the lower bound of equilibrium concentration levels still declines indefinitely with a growing market size, similar to the homogeneous goods case. However, with endogenous sunk costs, such as advertising and R&D expenditures, this property breaks down. If a firm can enhance the demand for its product at stage 2 by incurring, e.g., higher advertising costs at stage 1, such a decision will lead to higher sunk costs in the equilibrium. Since a growing market increases the achievable profits at the second stage, we may indeed expect the equilibrium level of sunk costs to be higher. Obviously, these higher sunk costs offset the tendency of an industry to become less concentrated as the market size increases. Hence, "*. . . a lower bound exists to the equilibrium level of concentration in the industry, no matter how large the market becomes.*" (Sutton, 1991, p. 11).

In Sutton (1998), the 'bounds' approach is applied to research-intensive industries by focusing on the relationship between an industry's R&D/sales ratio and its level of concentration. A crucial parameter in shaping this relationship is the 'escalation parameter', indicating the extent to which an industry consisting of many small firms can be destabilised by a firm by spending more on R&D than its rivals. Sutton shows that this (empirically unobservable) parameter sets a lower bound both to the maximal market share of a firm and to the R&D/sales ratio of the leading firm in an industry. Although the mechanism behind this resembles the endogenous sunk cost situation in Sutton (1991), a number of distinguishing features are worth mentioning here.

Given that a conventionally defined industry often captures several groups of products with different underlying technologies, complex overlapping patterns of substitutability on the demand side and of economies of scope (in R&D) on the supply side emerge. Since Sutton's aim is to develop an empirically testable framework, this complexity is explicitly acknowledged by modelling the R&D behaviour of firms as an allocation choice regarding their R&D projects. A firm can choose to focus on one particular technology (escalation) or to spread its R&D efforts across several technologies (proliferation). The more that R&D spending is effective in increasing the willingness-to-pay for the firm's products within the associated submarket, the more effective the escalation strategy will be. A high degree of substitution on the demand side between the product groups and the presence of scope economies in R&D will, however, make the escalation strategy more effective as well.

Sutton (1998) argues that in industries in which R&D spending is effective in raising consumers' willingness-to-pay, we can expect a high R&D/sales ratio. But the strength of the linkages between the submarkets eventually determines the level of concentration. For instance, if the degree of substitution on the demand side is low, the returns from escalation are low too, since it will only improve a firm's position in single, small submarket of the industry. However, the higher the degree of substitution, the more effective the escalation strategy will be. This automatically implies higher (minimal) levels of concentration, because any configuration in which the concentration level is lower than the minimal value will turn out to be unstable. In such a configuration, firms' fixed R&D expenditures must be small relative to the level of industrial sales. Under these circumstances it will be profitable for some firm to outspend its rivals on R&D, eventually resulting in a higher market share.

In conclusion, for those industries characterised by a high R&D/sales ratio, Sutton (1998) predicts that the lower bound to concentration should increase with the strength of the linkages between the submarkets. For industries exhibiting low R&D and advertising outlays, no lower bound to concentration exists. As in Sutton (1991), a rich set of statistical tests and detailed case studies accompany and support the theoretical framework by Sutton (1998).

3.2.3 Dynamic Equilibrium Models

Besides the static models mentioned above, also a number of dynamic equilibrium models have entered the industrial economic literature, mainly in the 1980s and 1990s. Their dynamic nature lies predominantly in their treatment of learning and selection as processes. Two of these models will still be shortly described here. The first one, Jovanovic (1982), is selected because of its pioneering way of modelling the entry process. The second model presented here is Ericson and Pakes (1995). Like Sutton's 'bounds' approach, this model is aimed at providing an empirical framework and could therefore serve well in explaining cross-sectional differences.

Jovanovic (1982) proposes a theory of 'noisy' selection. In his model of 'passive learning', firms have rational expectations with regard to distribution of true efficiency levels among the potential entrants. However, the most interesting assumption is that the firms do not know their (mean) individual level prior to entry. Only after entering, through receiving noisy information, does a firm discover its actual efficiency levels. If its actual efficiency exceeds its expectation, the firm will expand; in the other case it will contract or even exit the industry. The assumption that firms do not know their individual efficiency level prior to entry is interesting, because it is consistent with empirical observations regarding the productivity of entrants. Many firms enter with productivity levels far below the industry average (Baldwin, 1995), and consequently many of them leave the industry again shortly after entering. Another interesting feature of this model is that it explains the observation that smaller firms have higher and more variable growth rates than large firms. For explaining differences between industries, however, this model is under-determined, as it contains no industry-specific elements.

Ericson and Pakes (1995) propose a model that is aimed to serve as an empirical framework. Their model of 'active learning' is based upon a stochastic model of the entry and growth of a firm through the active exploration of its economic environment. This model assumes that a firm knows the current value of the parameter that determines the distribution of possible future streams, but that the value of this parameter changes over time in response to the stochastic outcomes of the firm's investments. Hence, in order to learn the value of a perceived opportunity, a firm must invest to enter the industry and to develop and (if feasible) exploit the opportunity. The stochastic outcome of this investment, together with the success of other firms in the industry and the competitive pressures from outside the industry, determine the profitability and the value of a firm. If the outcome is not favourable, a firm may find itself in an situation in which its idea is not perceived to be worth developing further and thus may decide to leave the industry.

This model of active learning is contrasted with a model a passive (Bayesian) learning in the empirical analysis of Ericson and Pakes (1998). As opposed to the active learning model, the passive learning model assumes that firms do not know the current value of the (time-invariant) profitability parameter. Only past profit realisations enable the firm to learn about the value of this parameter. Ericson and Pakes (1998) show that both models imply that the regression function of current firm size on initial size should be nondecreasing, but the passive learning model implies it should be strictly (i.e., monotonically) increasing, while the active learning model implies it need not be. Using longitudinal firm-level data, the authors conclude that the manufacturing data is consistent with the implications of the active learning model but is inconsistent with passive learning. The opposite conclusion holds for the data for retail trade. Unfortunately, the authors do not empirically exploit their data to search for differences *within* the manufacturing sector.

Although these and some other dynamic equilibrium models[32] allow for firm heterogeneity, technological uncertainty, and idiosyncratic (efficiency) shocks, still the level of rationality and forecasting abilities demanded from the agents are enormously high. For instance, in Jovanovic (1982), all (indefinitely small) firms have to know the entire equilibrium price sequence, whereas in Ericson an Pakes (1995) all firms are supposed to know their own and their competitors current efficiency levels, as well as the distribution of the industry's structure in future years.[33] As Dosi et al. (1995) argue, one is not at all sure what the dynamics would be if the rationality requirements of the agents were set much lower. Furthermore, they argue, it is difficult to accept that the speed of adjustment (by the agents) to changes in the selection environment is so high that empirically observed regularities are

[32]E.g., see Lippman and Rumelt (1982) and Hopenhayn (1992). More game-theoretical oriented models of industry dynamics can be found in, e.g., Maskin and Tirole (1988), Beggs and Klemperer (1992) and Rosenthal and Spady (1989). For a discussion of these models see Ericson and Pakes (1995).

[33]More formal criticism comes from Kaniovski (1998), who observes a central analytical inconsistency in Ericson and Pakes (1995), and from Dosi et al. (1995), who argue that the heterogeneity claim of models involving Markov processes should be taken with caution, because in the limit all agents visit infinitely often all attainable states and are therefore identical.

(near) equilibrium outcomes. Moreover, most of these models are difficult to translate into empirically testable models. Despite their dynamic approach, these models do not depict the properties of an industry for those periods in which it actually evolves into the predicted steady state solution. Only the properties of the steady state are depicted. Therefore, explaining cross-sectional differences by asserting that industries are in different stages of their evolution is not feasible with these models, basically because there is no explicit evolution.

3.2.4 Theoretical and Empirical Limitations of Equilibrium Models

During our discussion of the various classes of equilibrium models in the previous subsections, we have already indicated a number of problems associated with the empirical applications of these models. In what follows we aim to provide a more systematic overview of the theoretical and empirical limitations of the equilibrium models.

The most fundamental criticism of equilibrium models concerns the neglect of the process of change. Equilibrium models show the conditions for and properties of stable equilibrium configurations. Changes in the conditions underlying these configurations are analysed by comparative statics: the properties of the old equilibrium are compared with the properties of the new one. However, no reference is made to the way in which an industry moves from one equilibrium to another. In fact, disequilibrium cannot exist. This is an important shortcoming, because even if changes in the underlying conditions were discrete and only happened rarely, an industry may still be out of equilibrium for a considerable time, depending on the speed of adjustment. Only if all agents are fully rational and possess perfect forecasting abilities, they will instantaneously adjust to changing conditions.

However, behaviourists[34] have stressed that human beings are not perfectly rational. In many real-life situations, "*...decision problems are too complex to comprehend and therefore firms cannot maximize over the set of all conceivable alternatives. Relatively simple decision rules and procedures are used to guide action...*" (Nelson and Winter, 1982, p. 35). Therefore, whenever the equilibrium conditions of an industry change, the industry will be temporarily out of equilibrium as firms will need time to adapt to the new

[34]Most notably Herbert Simon; see Simon (1959, 1965).

situation. Again, if the equilibrium conditions seldom change, equilibrium analysis may still be appropriate, provided that the industry is in equilibrium most of the time. But technological and institutional changes, both from within the industry as well as from outside the industry, continuously alter the conditions underlying the equilibrium configurations.

Technological change, generally acknowledged to be a major determinant of economic growth, is especially difficult to embody properly in equilibrium models. The usual way of incorporating technological change in these models starts by assuming that technological change is exogenous to the economic system, and that, because of perfect information, the new knowledge associated with the innovation is immediately available to all firms. Given these assumptions, any innovation is immediately taken up by the firms and optimally[35] implemented in their production processes. As mentioned before, this necessarily implies that the relevant decision-makers within the firms know all the available production techniques and, equally important, how to apply them. Further, they will need to know the prices of all inputs, regardless of the scale on which these are hired. Finally, they will have to have the cognitive abilities to assess all the available techniques in order to select the one that minimises costs.

Obviously, this view on the innovative process has received a great deal of criticism. First of all, many firms devote resources to research and development in order to enhance their technological knowledge base. Since a large number of innovations are the outcome of these research and development activities, technological progress is endogenous rather than exogenous to the industry. Even in cases in which new technological knowledge becomes available to firms through research efforts from outside their industry, still they will have to undertake efforts to obtain and implement the knowledge. As such, the implementation of the exogenously available knowledge is an endogenous process as well.

This brings us to the second point of criticism, namely the diffusion of new technological knowledge. Equilibrium models assume that whenever an innovation is introduced in an industry, the knowledge related to the innovation flows around freely and competitors can thus rapidly incorporate the new knowledge. This assumption assures that any disequilibrium in the industry is only short-lived. However, it is plausible that both the tacit nature of technological knowledge and the bounded rationality of entrepreneurs prevent new knowledge from spilling over rapidly to competitors.

[35]In terms of cost minimisation.

As Nelson and Winter (1982) argue, a large component of technological knowledge is tacit, meaning that it cannot be articulated in codified forms such as blueprints or written instructions. But even if every new technology were fully specified in blueprints, bounded rationality would still limit the speed of diffusion. Entrepreneurs, like any other human being, do not have the cognitive powers to know and assess all the options available to them. Instead, the subjectively create a highly simplified model of their environment. Therefore, entrepreneurs vary in their perception of technological developments and their alertness to these developments. Together with the tacitness of knowledge this obviously hinders the rapid transfer of technological knowledge from innovators to imitators.

Because of the duration and imperfections of the knowledge diffusion process, innovative firms are able to enjoy at least temporarily higher (disequilibrium) profits. In fact, the opportunity to enjoy these profits provides the economic incentives to firms' search for technological improvements. But also in this respect the assumption of perfect information is obviously not plausible and even invalid. The innovative process involves some fundamental uncertainty, entailing the existence of techno-economic problems whose solution procedures are not known and the impossibility of precisely tracing consequences to actions (Dosi, 1988b). Consequently, entrepreneurs will sometimes make mistakes in this process or, at the minimum, behave in a way that is inconsistent with the assumption of rational profit maximisation. Of course, future profit opportunities may still be the main reason why firms engage in research and development, however a positive outcome of their search activities is never guaranteed *a priori*.

Since the outcomes of research and development programs are not known in advance, some firms will be more successful in the innovative process than others. Those firms that have successfully introduced an innovation increase their relative competitiveness vis-à-vis the unsuccessful ones. Competition between firms is therefore to a considerable extent endogenously driven by the *process* of technological change, continuously disturbing the economic status quo. Obviously, this rejects the notion of competition as a static concept. In conclusion, the inability to incorporate (technological) competition as a dynamic, perturbing process renders the equilibrium models largely inadequate for capturing the essence of real-life industry dynamics.

Besides these theoretical limitations of equilibrium models, a number of practical difficulties and limitations arise as well when translating the equilibrium approaches into empirically testable models. First of all, in some of these models the direction of causality is ambiguous. Especially the SCP-approach suffers from this endogeneity problem, where structure determines performance, but where performance may support some types of firm behaviour that could eventually affect market structure as well. A second 'empirical' problem with the class of equilibrium models relates to the uniqueness of the equilibria. In particular the game-theoretical models often depict multiple equilibria. Naturally, if a given model supports many outcomes, it is virtually impossible to find empirical support for its validity. Finally, some of the equilibrium models only focus on one dimension of industrial organisation. For instance, the 'bounds' approach mainly deals with explaining concentration levels. Explanations for, e.g., cross-sectional differences in market share turnover and profitability levels are unfortunately not captured by the 'bounds' approach.

3.3 TECHNOLOGICAL REGIMES

The framework of technological regimes significantly departs from the equilibrium models mentioned above, both theoretically as well as methodologically. From a theoretical point of view, the most important difference concerns the explicit recognition of competition as a technologically driven, dynamic process. Further, bounded rationality and technological uncertainty, mostly absent in equilibrium models, are important building blocks of the technological regime framework. Together, this makes the technological regime framework highly suitable for analysing the effects of technological search and learning processes.

The main difference in methodology between the equilibrium approaches and the technological regime framework can best be described by the distinction between formal and appreciative theorising, first brought forward by Nelson and Winter (1982). Formal theorising aims to set up a framework that enables one to explore, find and check logical connections between economic variables and to display possible causal mechanisms for particular phenomena. In formal theorising, analytical tractability is more important than minimising the intellectual distance from what is known empirically. Appreciative theorising tends to stay much closer to empirical substance. Here, theories are primarily conceptual tools of inquiry, and the focus is on the endeavour in which these tools are applied in explaining why certain economic events have happened. Although some overlap exists, it can be argued that formal theorising is the main methodological tool used in equilibrium models, whereas appreciative theorising much more underlies the technological regime framework.[36]

Given its theoretical and methodological foundations, the technological regime framework potentially serves well in providing realistic explanations for the observed differences between industries. In this section we will therefore elaborate on the technological regime framework. We will first focus on the evolution of the notion of technological regimes within the literature on economics and technological change. Attention will be paid to the different views on the role of technological opportunity and the role of knowledge in determining the rate and direction of technological change. Further, the role of institutional and market forces in selecting and establishing technological regimes will be considered. Second, we will present a short overview of empirical research on the relationship between technological regimes and the intersectoral variety in the rates and forms of organisation of innovations. Finally, we will examine the contribution of a number of evolutionary models to understanding the relationship between the structural and dynamic properties of an industry and its underlying technological regime.

3.3.1 Literature Overview

Not surprisingly, this overview of the literature on technological regimes starts with a discussion on the work of Joseph Schumpeter. During his life, Schumpeter held two rather different views of the innovative process. The first view, expressed in "The Theory of Economic Development" (Schumpeter, 1912), emphasises the role of new entrepreneurs in the innovative process. By introducing new ideas and innovations, new, small firms challenge the incumbent firms, contributing to what Schumpeter called the process of 'creative destruction' . However, the emergence of giant enterprises and industrial research laboratories during the first half of the twentieth century changed Schumpeter's view of the innovative process. In "Capitalism, Socialism and Democracy" (1942), Schumpeter calls attention to the key role of large firms as engines of economic progress by accumulating non-transferable knowledge in specific technological areas. Hence, his second view of the innovative process could be described as "creative

[36]For a discussion of the linkages between formal and appreciative theorising, see Nelson (1994).

accumulation" (Breschi et al., 1996). According to Freeman (1982), the main difference with Schumpeter's first view is: "... *the incorporation of endogenous scientific and technical activities conducted by large firms. There is a strong positive feedback loop from successful innovation to increased R&D activities setting up a 'virtuous' self-reinforcing circle leading to renewed impulses to increased market concentration.*"

During the 1950s, 1960s and 1970s only a few economists elaborated on Schumpeter's idea of putting technical change at the heart of the analysis of economic growth. Because of a preoccupation with unemployment and cyclical phenomena, unfamiliarity with natural sciences and technology, and major difficulties associated with the measurement of technical change, most economists preferred to consign technical change to a 'black box' (Coombs, 1988). The few economists that did concentrate on technical change mainly focused on particular innovations. With regard to the links between technical change and industrial organisation, causality was predominantly presumed to run from industrial structure (market concentration), through firm behaviour, to technological achievements, similar to the SCP approach. However, most of these studies[37] did not reveal this causality. Rather, there seemed to be a strong effect of industry-specific variables and a major influence of the degree of technological opportunity (at the industrial level) on the research intensity of the firms (Scherer, 1967, 1986; Kamien and Schwartz, 1982). Since a high research intensity creates technological entry barriers (in terms of high R&D investments required to compete successfully), concentration is likely to be affected by the research intensity of an industry and thus by the degree of technological opportunity. Therefore, a direct relationship exists between technological opportunity and industrial structure.

With regard to the link between technological opportunity and firm innovative behaviour, Griliches (1958) particularly focused on the diffusion process of inventions. Although in Griliches' analysis the technological opportunities are given, his contribution was important, since he was one of the first scholars to explain the intra-firm diffusion of inventions by economic forces. Once an invention is made, the diffusion of it (by means of firms managing the intra-firm adoption of the invention) can be explained in economic terms. Griliches (1958): "*Conceptually, the decisions made by an administrator of research funds are among the most difficult economic decisions to make and evaluate, but basically they are not very different from any other type of entrepreneurial decision.*"

---

[37]See Kamien and Schwartz (1982) for an extensive overview.

Schmookler (1966) went one step further by claiming that not only the diffusion of inventions, but also the pattern of inventive activity itself could be explained in economic terms, i.e., in terms of revenues and costs. According to Schmookler, the direction of inventive activities is mainly determined by demand considerations: the size of the market for specific inventions determines the composition of inventions. With regard to the supply side, he regards science and technology as an unlimited resource of inventions. Hence, from the supply side there are restrictions on neither the rate, nor the direction of inventive activities, and therefore it is demand that ultimately shapes this process.

According to Rosenberg (1976), the central weakness of Schmookler's demand-pull approach is the neglect of supply-responsiveness of technology and invention. His main objection is that inventions are not equally possible in all industries, because "*... there is a crucial intervening variable: the differential development of the state of subdisciplines of science and bodies of useful knowledge generally at any moment in time.*" He therefore argues that industries should be treated individually, since the supply of inventions depends on the knowledge bases upon which inventive activity in each industry can be drawn. But not only the rate of inventions is limited, inventive activities are also subject to what Rosenberg calls 'focussing devices'. These devices imply that inventive activities cannot explore in all directions, since "*... complex technologies create internal compulsions and pressures which, in turn, initiate exploratory activity in particular directions.*" (Rosenberg, 1976).

Nelson and Winter (1977) agree that, even under a wide range of demand conditions, some powerful intra project heuristics could apply that guide the evolution of certain technologies in a particular direction. They call these directions 'natural trajectories', and introduce the notion of a technological regime. A technological regime reflects the (perceived) boundaries of technological opportunities for further innovations and defines the trajectories to those boundaries. Like Rosenberg, they also recognise that underlying the movement along the natural trajectories involves "*... a certain knowledge on the part of the technicians, engineers, scientists, involved in the relevant inventive activity.*" (Nelson and Winter, 1977, p. 60). One key property of the knowledge base that may differ across industries is the extent to which scientific knowledge underpins the relevant technologies. That is, in some industries the role of science may be limited: the knowledge may then be quite specific and "*may involve more art and feel than science*" (Nelson and Winter, 1977, p. 60). However, in other industries science may play a major role in the process of technological change, providing generic knowledge relevant to the specific technologies.

Dosi (1982, 1988a) elaborates on both the natural trajectories and the characteristics of the knowledge base. To start with the latter, Dosi (1988a) distinguishes three aspects of knowledge: (1) the level of specificity, (2) the level of tacitness, and (3) the extent to which the knowledge is publicly available. The first aspect is quite similar to the property of the knowledge base put forward by Nelson and Winter (1977), i.e., knowledge can be universal and widely applicable, or more specific to particular 'ways of doing things'. The second aspect has to do with the extent to which the knowledge is well codified, or whether it is more tacit, learned mainly through practice. The third aspect relates to the fact that some knowledge is open and public, e.g., scientific and technical publications, whereas other knowledge is more private, either because it is protected by secrecy or patents which represents a choice behaviour of firms, or because it is tacit, which may either represent a choice not to 'articulate' the knowledge (akin to secrecy), or may be inherent in the process by which the knowledge needs to be exchanged or 'taught'.

With regard to Rosenberg's focussing devices, and Nelson and Winter's technological regimes and natural trajectories, Dosi (1982) adds to their concepts the notion of 'technological trajectories' and 'technological paradigms'. Similar to what a 'scientific paradigm' does in science in general[38], a technological paradigm sets out the relevant technological problems, and the pattern of solving these problems that are based on selected principles derived from natural sciences and on selected material technologies. A technological paradigm thus "*...embodies strong prescriptions on the directions of technical change to pursue and those to neglect.*" (Dosi, 1982, p. 152). These directions are then the technological trajectories, defined as the patterns of problem solving activities on the ground of a technological paradigm. Typical examples of technological paradigms and trajectories can be found in chemical processes, power generation, aircraft technology, microelectronic technology (Nelson and Winter, 1977; Saviotti and Metcalfe, 1984; Sahal, 1985), and in biotechnology (Orsenigo, 1989).

An interesting question posed by Dosi (1982) is what actually determines the emergence and establishment of a technological paradigm. Selection by market mechanism is not likely to be very strong *ex ante*, given the intrinsic uncertainty associated with the outcome of a technological paradigm.[39] Dosi (1982, p. 155): "*Other, more specific variables are likely to come into*

[38]See Kuhn (1962).
[39]See also Freeman (1974).

play such as (1) the economic interests of the organizations involved in R&D in these new technological areas, (2) their technological history, the fields of their expertise, etc.; (3) institutional variables 'strictu sensu' such as public agencies, the military, etc." The market is more selective *ex post* and *within* a technological paradigm. After a technological paradigm is selected and the technological trajectories have been defined, the market operates as (1) a selecting device among the range of products in which the technologies have been incorporated, and (2) as a signalling device, inducing producers to respond through technical advance, but mainly within the boundaries of the existing technological trajectory. New technological paradigms emerge "*...either in relation to new opportunities opened-up by scientific developments or to the increasing difficulty in going forward on a given technological directory...*" (Dosi, 1982, p. 157).

Dosi (1988a) suggests that the empirically observed inter-sectoral differences in the rates and forms of organisation of innovations can be explained by differences in paradigm-specific opportunities.[40] Since opportunities for innovation depend on the stage of development of a technological paradigm (by definition, the establishment of a new paradigm increases both the scope and the ease with which potential innovations can be achieved), unevenly distributed degrees of development across sectors can explain the inter-sectoral differences with regard to innovations.

However, there are also differences *between* technological paradigms. One aspect in which technological paradigms differ (and, related, the opportunities for innovation) has already been mentioned, i.e., the underlying properties of the knowledge base. But differences in the paradigm-specific opportunities may also stem from "*... (1) the 'ease' with which technological advances, however defined, can be achieved; (2) different possibilities for the innovator to appropriate economic benefits from it in terms of profits, market shares, etc.; and (3) different degrees of cumulativeness of technological advances...*" (Dosi, 1988b, p. 233). The 'ease' with which technological advances can be achieved in a certain sector depends on, for instance, the extent to which that sector can draw from the knowledge base and the technological advances of its suppliers and customers. Appropriability refers to the possibility of innovators to reap the profits from their innovations and to protect it from imitation. Possible appropriability devices[41] are patents, secrecy, lead times, costs and time required for duplication, learning

[40]Scherer (1967, 1986) and Kamien and Schwartz (1982) provide similar suggestions.
[41]See Levin et al. (1984).

curve effects, superior sales efforts, and differential technical efficiency due to scale economies. Several studies[42] have shown that the combination of these devices, as well as the overall degree of appropriability, differs between sectors. Finally, dynamic returns to innovative effort or auto-correlated probabilities of innovative success, indicating the cumulativeness of technological advance[43], may differ between sector because of differences in the cognitive nature of the learning process (e.g., learning by doing). It depends on the extent to which the underlying technology is cumulative, i.e., the extent to which "*... today's technical advances build from and improve upon the technology that was available at the start of the period, and tomorrow's in turn builds on today's.*" (Nelson, 1995, p. 74).

For a clear understanding of the differences between technological paradigms and technological regimes, it is perhaps useful to express our view on the distinction between the two concepts. A *technological paradigm* defines the technological opportunities for further innovations and channels the direction of technological activities within a paradigm (i.e., the technological trajectories). The emergence and establishment of a technological paradigm is mainly determined by institutional factors, whereas market selection is more important within a paradigm (i.e., along the technological trajectories). Finally, each technological paradigm is characterised by a particular combination of the following elements: (1) the 'ease' of achieving technological advances, (2) the level of appropriability, (3) the level of cumulativeness, and (4) the properties of the knowledge base, common to specific activities of innovation and production and shared by the population of firms undertaking those activities.

This particular combination of opportunity, appropriability, cumulativeness conditions and properties of the knowledge base in turn determines the *technological regime*, which specifies the economic logic governing the introduction and adoption of the paradigm-specific technologies. Hence, a technological paradigm defines the opportunities of the innovative process mainly in technological terms, whereas a technological regime consists of the economic opportunities and constraints.

To explain how a technological regime specifies the economic logic governing the introduction and adoption of paradigm-specific technologies, usually two combinations of opposite regime conditions are analysed in the

[42] See Dosi (1988a) for an overview of these studies.

[43] See also Teece (1982, 1986).

literature on technological regimes. The first one is characterised by low appropriability and cumulativeness conditions, and the knowledge is mainly specific, codified and simple.[44] All these conditions facilitate innovative activities by young and small firms. Low appropriability conditions impede the persistence of high profits associated with previous innovations and, as a consequence, they impose a barrier to growth to previously successful innovators. And because cumulativeness is low, previously successful innovators cannot benefit from dynamic increasing returns to their innovative efforts. Finally, the properties of the knowledge base contribute to potentially high spillover levels. When the knowledge underlying an innovation is specific[45], codified and simple, it spills over rather easily, which facilitates imitation by other firms.

Therefore, when appropriability and cumulativeness conditions are low, and the knowledge is mainly specific, codified and simple we can expect the pattern of innovative activities to be characterised by the constant inflow of new entrepreneurs with new ideas, products and processes, continuously challenging the established firms and wiping out their quasi rents associated with previous innovations. Since such a pattern resembles the first one of Schumpeter's two views on the innovative process (as set out at the beginning of this subsection), a regime with such underlying conditions is often called a Schumpeter Mark I (SM-I) regime.

This SM-I regime is contrasted with a technological regime where the three conditions are reversed. Here, appropriability and cumulativeness conditions are high, whereas the knowledge is mainly generic, tacit and complex. Under these conditions, the patterns of innovative activities are likely to be dominated by large and established firms. High appropriability conditions facilitate the persistence of high profits associated with previous innovations. Therefore, previously successful innovators will have sufficient financial means to expand their operations. Further, because of high cumulativeness the previously successful innovators can explore dynamic increasing returns to their innovative efforts, which reinforces their innovative successes. Finally, the properties of the knowledge base impede other firms to imitate successful innovators, because generic, tacit and complex knowledge by definition implies low spillover levels.

[44]Opportunity conditions are assumed to be similar in these two regimes.

[45]In the sense that generically applicable scientific knowledge plays a minor role.

Since the pattern of innovative activities resulting from such conditions is likely to be characterised by the dominance of large and established firms and by high (technological) entry barriers, this regime closely resembles Schumpeter's second view on the innovative process. Therefore, a regime with high appropriability and cumulativeness conditions and with generic, tacit and complex knowledge conditions is called a Schumpeter Mark II (SM-II) regime.

In conclusion, the conditions defining a technological regime provide a set of economic opportunities and constraints for the population of innovative firms. As such, the technological regime framework provides a plausible theoretical explanation for the inter-sectoral variety in the rates and forms of organisation of innovations. To what extent the empirical literature on technological regimes supports this explanation will be investigated in the next subsection.

3.3.2 Empirical Evidence on Innovative Patterns

The most important empirical work in this area can be found in Malerba et al. (1995) and Breschi et al. (1996). By using patent data for four countries (Germany, France, United Kingdom and Italy) for the period 1968–1986, Malerba et al. (1995) construct a group of indicators of the Schumpeter Mark I and Schumpeter Mark II patterns of innovative activities, and call these patterns respectively 'widening' and 'deepening'. Next, they group the empirical proxies into four types: concentration of innovative activities, size of the innovating firms, change over time in the hierarchy of innovators, and the relevance of new innovators. After applying these indicators to 33 technological classes, the authors find the following results.

First, patterns of innovative activities differ systematically across technological classes. Second, they observe that across countries, remarkable similarities emerge in the patterns of innovative activities for each technological class. In the same technological class the indicators approximately have the same values across countries. Third, two groups of technological classes can be defined, in which innovative activities are organised according to the widening and the deepening patterns of innovative activities. In general, mechanical and traditional sectors show a widening pattern, whereas chemicals and electronics better show a deepening pattern. Especially the second result "*...suggests strongly that 'technological imperatives'* *and technology-specific factors (closely linked to technological regimes) play a major role in determining the patterns of innovative activities . . .*" (Malerba et al., 1995, p. 47).[46]

Breschi et al. (1996) estimate the impact of a number of technological regime variables on Schumpeterian patterns of innovation (defined by the specific combination of entry, stability and concentration of the innovating firms). They find considerable evidence for the existence of a relationship between sectoral patterns of technical change and the nature of the underlying technological regime. Especially the deepening patterns of innovative activities seem to be related to high degrees of cumulativeness and appropriability, and high importance of basic sciences relative to applied sciences. Another interesting conclusion is that widening patterns of innovation are associated to either very high or very low opportunity conditions. The authors explain this result by the ambiguous effects of opportunity conditions on concentration. High levels of opportunity facilitate the entry of new innovative firms, which leads to an increase in the population of innovation and thus lower concentration levels. But also low opportunities can have a negative effect on concentration by impeding the persistence of major differences in innovative rates among firms.

In conclusion, both of these papers provide evidence for the hypothesis that a relation exists between the pattern of innovative activities in a sector and its underlying technological regime. However, although the previously mentioned authors have used some demographic indicators, they have only employed these indicators to the population of firms registered in patent databases. The question that naturally follows concerns whether a similar relationship exists between the structural and dynamic properties of the full population of firms in an industry and its underlying technological regime.

3.3.3 Technological Regimes and Industrial Dynamics in Evolutionary Models

Such a relationship, running in first approximation from the latter to the former, has indeed been suggested by, e.g., Nelson and Winter (1982), Winter (1984) and Dosi et al. (1995). Nelson and Winter (1982) have developed a number of models that capture some elements of technological

[46]Malerba et al. (1996) perform a similar analysis with a different dataset, however the results are "*remarkably consistent*".

regimes. Without going into details, the main results derived from the simulations of these models are the following. In general, higher opportunities for innovation (parameterised by a higher growth of latent productivity, i.e., the movement of the set of new technological possibilities created outside the industry) lead to higher concentration levels. The model's explanation for this is that if one firm is successful in R&D, it will create a larger advantage of that firm over non-successful innovators in case of high latent productivity growth.

Being successful in R&D is modelled as a random process, in which a firm receives a draw from a distribution of technological opportunities. Since higher latent productivity growth is reflected by the distribution of opportunities, the 'productivity jump' of a successful innovator will therefore be higher when latent productivity growth is high. This, in turn, leads to a higher expansion of the firm (in the model, net investments are a function of price-cost margin, which is positively related to productivity levels). Another important result is that a lower ease of imitation (implying higher appropriability) leads to a higher ultimate concentration level of the industry. The explanation is quite straightforward: if it is hard for firms to imitate the technology of an innovator, the latter will be able to reap the benefits from its innovation longer than in case of easy imitation, and will therefore expand more rapidly.[47]

Nelson and Winter (1982) admit that the exclusion of entry of new firms in their models might be an important omission. Winter (1984) overcomes this omission by presenting an extension of one of the models presented in Nelson and Winter (1982), in which the role of innovative entry on industry evolution is examined under two different technological regimes. The first regime is labelled 'entrepreneurial', and resembles more or less Schumpeter's earlier view of the innovative process, as expressed in "The Theory of Economic Development". Schumpeter's later view, expressed in "Capitalism, Socialism and Democracy" underlies the 'routinised' technological regime. These different regimes are represented by different parameter settings[48], such that an entrepreneurial regime is favourable to innovative entry and unfavourable to innovative activity by established firms. In a routinised regime these conditions are reversed, of course.

[47]Both this and the previous result were found if the industry was relatively unconcentrated initially. Industries that were initially already concentrated were found to remain very stable with regard to concentration, regardless of the parameter setting.

[48]More specifically, the parameters governing the rate of innovative entry and the productivity level resulting from an innovation have different values between the two regimes.

The two most important parameters are the relative importance of externally and internally generated innovations, and the degree of comprehensiveness of a single innovation. Externally generated innovations, which are more important in the entrepreneurial regime, are the result of 'background' R&D activity that is relevant to the industry's technology but is not funded by the industry itself (as opposed to internally generated innovations). Hence, this first parameter basically reflects differences with respect to properties of the knowledge base relevant to the industry. The second parameter relates to cumulativeness conditions. In the entrepreneurial regime, an innovation is comprehensive in the sense that if a firm is successful in innovating, the firm's new productivity level is equal to the productivity level associated with the new technique it found. In the routinised regime, however, the new productivity level is the geometric mean of the level associated with the innovation and the level associated with the old technique. Including some additional assumptions with regard to the productivity growth of the old technique, actual innovative entry is tended to be confined to the early stages of the industry's development in the routinised regime. In later stages, "*... the potential innovative entrant faces the handicap that the single good idea represented by his one non-comprehensive innovation tends not to be enough to permit him to match the efficiency of the established competition ...*" (Winter, 1984, p. 221).

The simulation runs show a number of interesting results. With regard to the ratio of adopted innovations associated with new entry and the innovations by established firms, the simulations clearly show that in the entrepreneurial regime the majority of innovations arise from the entry of new firms, as opposed to the routinised regime. Further, the total number of innovations in the routinised case is much higher than in the entrepreneurial regime, which offsets their more incremental nature. Hence, by the end of the run productivity is substantially higher in the routinised case. Finally, in the entrepreneurial regime the industry structure is much less concentrated and less profitable at the end of the run than in the routinised regime.

The final model discussed here is by Dosi et al. (1995). In this model, technological regimes (together with 'market regimes', indicating the relative speed at which selection occurs) are used to explain the empirically observed intersectoral variety in the industrial structures and dynamics. In the model, each technological regime is defined by the stochastic processes determining the 'competitiveness' of entrants and incumbents. For instance, in one regime, called Schumpeter Mark I, the competitiveness of incumbents remains the one they were endowed with at their birth. Hence, in this regime they never learn, as opposed to the Schumpeter Mark II regime. Here, learning by incumbents is highly cumulative: in each period the competitiveness of incumbents is stochastically augmented by a factor that is positively related to their relative competitiveness (i.e., *vis-à-vis*the average competitiveness of the industry). Thus, in Schumpeter II regimes, 'success breeds success'. Finally, an Intermediate Regime is defined, in which the competitiveness of incumbents essentially follows a random walk with a positive drift.

Interestingly, the simulation results show that within a Schumpeter I regime some intertemporally stable, but lowly concentrated, structures emerge. This is not the case for the other two regimes that display an irregular long-term cyclical pattern. In a Schumpeter II regime, however, the average time that the industries spend in highly concentrated structure is longer than in an Intermediate regime. Hence, disruptions occur less frequently under Schumpeter II, but they induce major discontinuities when they occur. With regard to firm size distributions, the simulations show that the less skewed size distribution appear under Schumpeter I regimes, whereas the most skewed appear under Schumpeter II.

Besides these comparisons between the three regimes, the experiments by Dosi et al. (1995) also allow for an analysis of different parameter settings within the regimes. Three parameters are considered here: the technological opportunities for incumbents, the technological opportunities for entrants, and the level of market selection. In general, higher opportunities for incumbents lead to a smaller population of firms, higher concentration, and lower market turbulence.[49] Higher technological opportunities for entrants lead to exactly opposite results: a higher number of firms, lower concentration, and higher market turbulence. Finally, a higher level of market selection has a negative effect on the number of firms, and a positive effect on concentration.[50] Hence, higher selective market forces (more 'efficient' markets) "*. . . tend to yield, in evolutionary environments, more concentrated market structures, rather than more 'perfect' ones in the standard sense.*" (Dosi et al., 1995, p. 13).

3.3.4 Conclusion

This section has shown that the technological regime approach provides a sound framework for explaining the inter-sectoral variety in the rates and

[49]Measured as the sum of the absolute changes in market shares.

[50]No clear relationship was found between market selection and turbulence.

forms of organisation of innovative activities. By explicitly recognising competition as a dynamic process, and by allowing for bounded rationality and technological uncertainty, the theoretical foundations of the technological framework are based on very realistic assumptions. Given these assumptions, the technological regime frameworks explains the sectoral asymmetries in innovative activities by differences in opportunity, appropriability, cumulativeness and knowledge conditions underlying the innovative activities. As we have shown, this explanation is supported by substantial evidence from empirical research that focussed on the structure and dynamics of the *population of innovating firms*.

The evolutionary models discussed in the previous subsection also depict a theoretical relationship between the structural and dynamic properties of the *full population of firms in an industry* and its underlying technological regime. In chapter 4 we will test whether such a relationship can also be established empirically. However, differences in the structural and dynamic patterns between industries may also stem from the fact that the industries are at different stages of their life cycle. In the next section we will elaborate on how theories on product or industry life cycles depict the evolution of industries.

3.4 PRODUCT LIFE CYCLES

The product life cycle approach starts from the observation that most successful products exhibit a life cycle that basically consists of four subsequent stages: introduction, expansion, consolidation, and contraction. When the product is introduced, sales are low and prices are high, but in the expanding stage sales increase rapidly and prices fall significantly. Both trends level off in the consolidating stage, after which sales start to fall during the contracting stage; for prices the trend is less clear during this stage. These regularities in itself may have important implications for firms marketing their products[51], but from the perspective of this chapter, the most interesting aspects of the product life cycle are the co-evolving structural and dynamic properties of the industry. This section will present the most important contributions to the literature on product or industry life cycles.

[51] See e.g., Dean (1950), Levitt (1965), or Hayes and Wheelwright (1979a and 1979b).

3.4.1 Literature Overview

In the late thirties, Joseph Schumpeter[52] already observed that after the introduction of a major product innovation, a strong 'band-wagon' effect often appears, followed by the entry of many new firms into the rapidly expanding sectors, attracted by the high profits associated with innovations. However, the entry of new firms leads to an erosion of profit rates, and as the industry matures and its product becomes standardised, a shift to cost-saving innovations occurs and the exploitation of scale economies becomes more important, leading to a reduction in the number of firms in the industry. In the subsequent decades, a number of detailed industry studies[53] have confirmed and refined these regularities. For instance, the evolutionary patterns of automobiles, car tires, televisions and television picture tubes, penicillin, typewriters, and commercial aircraft for trunk carriers all showed a sharp shakeout, together with a peak of entry rates in the initial build-up of the industry and negligible entry shortly after the shakeout. Furthermore, these studies observed that market shares tend to stabilise over time, with a general dominance of the earliest entrants in these industries (Klepper, 1997).

Textbook models of perfect and monopolistic competition are not able to reproduce these regularities. The main reason for this is that such models are static, whereas the product life cycle is essentially a dynamic process. But even applying comparative static analysis could not show the patterns described by the product life cycle framework. Suppose the product life cycle would be simulated in a model of perfect competition by an exogenously shifting demand curve. In such models the relationship between demand and the number of firms would always be positive and linear: ceteris paribus, the change in the number of firms is always equal to the total change in demand divided by the cost minimising size of the representative firm.

In the framework of the product life cycle, however, this relationship is not so straightforward. There can even be an increase in demand and a simultaneous reduction in the number of firms in the industry. This last regularity can be reproduced only by an (again) exogenously changing production function such that the average cost minimising size of the representative firm increases. Still, this would not reflect the empirically observed gradual standardisation of the industry's product over its life cycle,

[52]See Schumpeter (1939).

[53]See Klepper (1997), and Klepper and Simons (1999) for a detailed overview of these studies.

simply because in models of perfect competition the products are assumed to be homogeneous. Models of monopolistic competition could reproduce this result by imposing a decreasing elasticity of the firm's individual demand curve. But then again the adjustment needed to reproduce the different regularities of the different stages of the product life cycle would have to be imposed on the model.

In conclusion, static models of industrial organisation require arbitrary assumptions to reproduce the regularities of the different evolutionary stages of an industry by exogenously changing the relevant parameters of the model. But these models cannot explain why these parameters change over the life cycle of an industry. An approach for modelling and analysing the relationships between the product life cycle, the co-evolving nature of technological change and their impact on industrial dynamics that does not require such arbitrary assumptions would obviously be much more satisfactory.

In line with Klepper and Simons (1999), we divide the most important contributions in modelling and analysing these relationships[54] in two 'schools'. The first one, labelled 'event theories' presumes a well-defined technological event or development that increases the value of experience of some firms. The second school, called 'competitive advantage theories', emphasises firm heterogeneity coupled with intensifying competition over time. Both approaches show how a shakeout of firms can take place in the evolution of the industry. We will first discuss the event theories.

The first important contributions in the event theories have been provided by William Abernathy and James Utterback[55], by introducing the notion of a dominant design. In their view, new industries are initially characterised by uncertainty about the product's technology and the user preferences. In this 'experimental' stage market shares often change rapidly between successful innovators and less efficient rivals, with firms mainly competing on product innovations. However, as both producers and consumers gradually select the desired features of the product, the opportunities for improving the product decrease. At this stage a 'dominant design' emerges that sets the standards for the product. "*A dominant design . . . is synthesized from more*

---

[54]Naturally, the product life cycle concept has also been studied by other economic disciplines. We have mentioned already some contributions from the business literature (see footnote 27), but also in the international trade literature the product life cycle has been employed; see Vernon (1966, 1979). However, we will only discuss the literature on product life cycles in industrial economics.

[55]See, for instance, Utterback and Abernathy (1975) and Abernathy and Utterback (1978).

fragmented technological innovations introduced independently in prior products and tested and often modified by users of those prior products." (Suárez and Utterback, 1995, p. 418) Typical examples of emerging dominant designs can be found in the typewriter industry, automobiles, television and picture tubes and the transistor industry.

Obviously, the technological event here is the emergence of a dominant design that shifts the terms of technological competition in an industry. In this stage, firm survival is largely determined by the ability to adapt to the new technological environment. Firms that are not able to efficiently produce the dominant design are forced to exit. Further, entry and survival rates decline too, because (1) the opportunities for new product designs are depleted, and (2) the experience, knowledge and reputation build up by earlier entrants have become important competitive assets, reducing the relative competitiveness of the new entrants and thus decreasing their survival chances. Together with the increasing exit of firms not sufficiently adapting to the new technological environment, this leads to a shakeout of firms in the industry.

After the emergence of a dominant design, exit still continues, as surviving firms gradually shift their attention from product improvements to means of producing more efficiently by investing in capital-intensive production methods. This increases the minimum efficient size of firms in the industry, which forces firms below this size to exit as well. Eventually, the industry consolidates as the opportunities for improving the production process are depleted. Hence, process innovations occur less frequently and become more incremental, causing market shares to stabilise.

An important condition for the emergence of a single dominant product design is that the consumers of the product have similar demands. Obviously, not all markets satisfy this condition. Windrum and Birchenhall (1998) provide three counterexamples of mature markets characterised by divergent consumer needs (camera, musical amplifier and personal computer markets) which all have differentiated into a number of market niches. Moreover, they observe, even within these niches alternative designs may compete. For instance, within the group of professional photographers, a variety of film formats, lenses and shutter mechanisms are being used. In their simulation model, embodying coevolutionary, interpopulation dynamics of consumer and firm learning, they aim to analyse under what conditions such a pattern of multiple niches is likely to emerge. The authors' simulation experiments are based on differences in the initial number of firms to consumer types. The results show that in most of the simulation runs the market converges to more than one consumer type serviced by multiple firms. Only in some cases a single dominant design emerges. The likelihood of this event decreases as the initial number of firms to consumer types increases. Hence, the authors conclude that the emergence of a single dominant design is not more than a 'special case'.

Within the event theories, the model by Jovanovic and MacDonald (1994) offers an alternative technological event that shifts the terms of competition in the industry. Instead of the emergence of a dominant design, a major invention emanating from outside the industry eventually causes the shakeout of firm. The intuition behind their model is as follows. At the moment the invention arrives, a group of firms trying to exploit the invention will enter the industry. Both incumbents and entrants will try to implement the invention, but only the successful ones will be able to decrease their costs and expand to a greater optimal size. Being successful is determined by a random process, in which the more experienced incumbents have a higher probability to implement the innovation than the post-invention entrants. The output growth and declining prices resulting from the expansion of successful firms render technological laggards unprofitable and force them to exit the industry. Since entry does not occur after the introduction of the invention, the exit of the laggards constitutes the shakeout in their model.

In the comparative advantage theories, shakeouts are due to a process of continuously increasing competition instead of the emergence of a dominant design or a radical invention. Within these theories, the organisational ecology approach focuses on non-technological competition. Organisational ecology[56] primarily studies the evolution of heterogeneous populations of organisational forms, but it departs from the traditional organisational theories by focussing much more on the population of organisational forms rather than on the behaviour of individual organisations. An important assumption in this theory is that organisations are relatively inert, i.e., they adapt rather slow to changes in the selection environment. Competition (for scarce resources) and legitimation (the social acceptance of an organisational form) determine the evolutionary pattern of populations. The strength of both these forces depends on the density of the population in the following way. The closer the number of organisations is to the maximum number of organisations supported by the resource environment (the so-called carrying capacity), the more intensive the selective forces will be.

---

[56]Hannan and Freeman (1989) and Hannan and Carroll (1992) are the basic references for studies in this field.

The most interesting aspect from the point of view of this book is the predicted relationship between density on the one hand, and founding and survival rates of organisational forms on the other. In the early stage of the evolution of a population, founding and survival rates will be low due to the absence of sufficient legitimacy. But as legitimacy increases over time, both founding and survival rates will grow up to a point where the carrying capacity of the environment is approached. At this point the increased competition for the scarce resources will cause founding and survival rates to decline again.

As with the product life cycle theories, organisational ecology could explain differences between industries by asserting that industries are in different stages of their evolution. However, we have not included the organisational ecology approach in our empirical analysis for the following reason. The organisational ecology approach strongly focuses on the emergence and coexistence of different organisation forms of firms, but with the available data it is virtually impossible to distinguish these different organisational forms. Therefore, despite the focus on non-equilibrium interactions and the recognition of heterogeneity and bounded rationality, we have not explored the empirical opportunities of the organisational ecology approach.

Comparative advantage theories embodying increasing technological competition can be found in Gort and Klepper (1982), Klepper and Graddy (1990), and Klepper (1996). Gort and Klepper construct a theory of the evolution of a new industry to explain the history of the diffusion of 46 products. They distinguish five evolutionary stages with respect to the number of producers in a prototypical new industry. The commercial introduction of a new product marks the beginning of stage I. Stage II is defined as the period of sharp increase in the number of producers, whereas in stage III the number of entrants more or less balances the number of exiting firms. Stage IV then is the period of negative net entry, and the final stage V is the second period of approximately zero net entry. However, Gort and Klepper do not consider the first and the final stage and thus apply their following model only to stages II, III, and IV.

By definition, in every period the expected number of entrants is equal to the probability of entry of each potential entrant, multiplied by the number of potential entrants. The probability of entry depends on information on new product technology, including the potential rewards of entry. The information stems from two sources: (1) non-transferable information emanating from experience in production by existing firms, and (2) information emanating from sources outside the set of current producers. The first type of information has a negative effect on the probability of entry, since it provides an advantage of incumbents over entrants and hence operates as an entry barrier. Because the second type of information reduces the value of accumulated experience of incumbents, it has a positive effect on entry. In the early stages of the product life cycle, most innovations are based on the second type of technological information, whereas at later stages the innovations emanate more from the accumulated stock of experience of existing firms.

Therefore, entry rates are high in the early stage of an industry, when the number of firms is relatively low and, consequently, profits are high. But as the industry matures, three forces reduce the entry of firms: (1) the accumulation of experience by existing firms, (2) the decrease in profit rates resulting from an increase in the number of producers, and (3) a gradual reduction in the population of potential entrants that have not yet entered the market. Eventually a point of zero net entry is reached, but as the above mentioned process continues, entry rates are still falling and exit rates will rise sharply, because the increasing competition compresses profit margins and forces less efficient firms to leave the market. Gort and Klepper (1982) find substantial support for their hypotheses in analysing the 46 product histories. However, as they state in their conclusion: "*We view many of these inferences as only first steps toward developing a theory of the evolution of industries.*" (Gort and Klepper, 1982, p. 651).

In this respect, Klepper and Graddy (1990) could be seen as a second step in developing such a comparative advantage theory. After re-analysing the data set assembled by Gort and Klepper, they establish three evolutionary stages, again with respect to the number of firms in a new industry: first it grows, then declines sharply, and finally it levels off.[57] Based on this and other inferences about output and prices, Klepper and Graddy construct a model, which aims at explaining the observed regularities. In comparison with Gort and Klepper (1982), the model by Klepper and Graddy focuses more on individual firms. They assume that the limited number of potential entrants differ in the average costs and product quality, and that cost reductions only occur immediately after entry by means of imitating more efficient rivals. All firms in the industry change their capacity in each period at a rate that is determined by their average cost relative to the prices.

[57]Hence, these stages resemble respectively stages II, IV and V of Gort and Klepper (1982).

Including some other assumptions, Klepper and Graddy (1990) show that their model is "*capable of explaining the fall in price, the rise in output, and the eventual levelling off of both that characterizes the evolution of new industries.*" (Klepper and Graddy, 1990, p. 39). According to the model, in the early stage of a new industry, when prices are relatively high, the number of firms increases because all firms are profitable, and consequently no prior entrants exit the industry. But as the price decreases, firms with high cost levels will eventually make losses and therefore exit the industry. At some point, the price will attain a level at which profitable entry is no longer possible, causing entry to cease. With exit still continuing, the number of firms in the industry drops significantly, until eventually price levels off, which stabilises the number of firms.

Although the model by Klepper and Graddy is capable of reproducing many empirical regularities concerning the evolution of new industries, the authors admit that their model is highly simplified, excluding factors as innovation by incumbents, scale economies, and strategic behaviour. Further, the gradual shift over time from product innovations to process innovations, as observed by for instance Schumpeter (1939), is absent in their model as well. The model by Klepper (1996) does include many of these issues. Furthermore, it allows for the derivation of a number of testable hypotheses, as will be shown in the next chapter. For these reasons we believe that Klepper (1996) provides the most appropriate conceptual tool to study empirically the relation between the evolutionary stages of industries and their structural and dynamic properties. The next subsection will highlight the most important features of this model.

3.4.2 Klepper's Model

The basic idea of Klepper's model is to depict the evolution of a new, technologically progressive industry. In this model, all the decisions of the firms are made to maximise current expected profits. For firm i in period t, the expected profit $E(\Pi_{it})$ can be expressed as:

$$E(\Pi_{it}) = [s_i + g(\mathrm{rd}_{it})]G - F - \mathrm{rd}_{it} + [Q_{it-1}(Q_t/Q_{t-1}) + \Delta q_{it}] \\ \times [p_t - c + l(\mathrm{rc}_{it})] - \mathrm{rc}_{it} - m(\Delta q_{it}) \quad (3.1)$$

In the first part of this expression, $[s_i + g(\mathrm{rd}_{it})]$ is the probability of firm i developing a product innovation. This probability is determined by two factors. The first one is the firm's innovative expertise s_i, with which it is randomly endowed at its birth and which remains constant over time.[58] The second factor is the firm's spending on product R&D, i.e., rd_{it}. The function $g(rd_{it})$ reflects the firm-specific opportunities for product innovation. The following conditions apply:

(i) $g'(rd_{it}) > 0, \ \forall rd_{it} \geqslant 0$
(ii) $g''(rd_{it}) < 0, \ \forall rd_{it} \geqslant 0$
(iii) $g'(0)\ G > 1$
(iv) $\forall t, \ \exists i$ s.t. $s_i = s_{max}$
(v) $s_i + g(rd_{it}) \leqslant 1$

G denotes the one-period gross monopoly profit from offering a distinct product variant achieved by a successful product innovation. The innovators sells the distinctive variant of the product at a higher price than the price of the standard product, but the monopoly profits are only made for one period, because other firms monitor all the innovations at some fixed costs[59] F and subsequently imitate the innovation without any additional costs. After one period, all product innovations are copied and incorporated in the standard product. It is assumed that $F > [s_i + g(rd_{it})]G - rd_{it}$ for all rd_{it}. Hence, all firms must produce the standard product in order to have nonnegative profits. Condition (iii) ensures that $rd_{it} > 0$ for all i, t.

The second part of expression (3.1) denotes the firm's net profit from producing the standard product after subtracting its spending on process R&D, rc_{it}, and the costs of adjusting its output, $m(\Delta q_{it})$. Q_t, the total quantity demanded, expands over time as the price of the standard product, p_t, falls over time. A key assumption is that incumbents can maintain their market share at no additional costs. Therefore, they experience a rise in their sales from Q_{it-1} in period $t-1$ to $Q_{it-1}(Q_t/Q_{t-1})$ in period t, where Q_t/Q_{t-1} is the growth in total quantity demanded. If it wants to increase its output further (i.e., increasing its market share), it must incur an adjustment cost of $m(\Delta q_{it})$, where Δq_{it} is the expansion in its output above $Q_{it-1}(Q_t/Q_{t-1})$. The function $m(\Delta q_{it})$ is such that $m'(0) = 0$, $m'(\Delta q_{it}) > 0$ for all $\Delta q_{it} > 0$, and $m''(\Delta q_{it}) > 0$ for all $\Delta q_{it} \geqslant 0$. Hence, increasing market share is assumed to be subject to increasing marginal adjustment costs.

[58]The cumulative distribution function $H(s)$, from which the innovation expertise s_i is drawn, is the same for the potential entrants in each period.

[59]According to Klepper (1996), these monitoring costs F have to be incurred by a firm "*in order to be able to imitate costlessly the innovations of its rivals, which is required to market a distinctive product variant and also the standard product....*"

The function $l(\text{rc}_{\text{it}})$, reflecting the opportunities for process R&D, is assumed to asymptotically approach an upper bound as rc_{it} increases. Furthermore it reflects diminishing returns. Hence, $l'(\text{rc}_{\text{it}}) > 0$ and $l''(\text{rc}_{\text{it}}) < 0$ for all $\text{rc}_{\text{it}} \geqslant 0$. Because of the assumption that $l'(0)\ Q_{\text{min}} > 1$, where Q_{min} is the smallest level of output ever produced by any firm, $\text{rc}_{\text{it}} > 0$ for all i, t. Finally, the average cost of producing the standard product[60] is assumed to be independent of Q_{it} and equal to $c - l(\text{rc}_{\text{it}})$. Thus if no firm spends money on process R&D, the average cost of producing the standard product is equal to c, and it is the same for every firm.

If the expected profits are negative, incumbents will exit the industry. Entrants will only enter if the expected profits are positive. In each period t there is a limited number of potential entrants K_t. For entrants, expression (3.1) applies as well, with $Q_{\text{it}-1}$ equal to zero. It is assumed that all firms are atomistic and price takers, and the market is cleared in every period. In every period firms decide on how much will be spent on product R&D, process R&D, and the expansion in market share to maximise current expected profits. As we have discussed in section 3.2.4, this is a strong assumption, that conflicts with the more realistic assumption of bounded rationality. On the other hand, the behavioural assumptions of Klepper (1996) are much less strong than the assumptions underlying the dynamics equilibrium models as discussed in section 3.2.3, because the agents in Klepper's model are assumed to be myopic instead of knowing all future states of the industry.

In Klepper (1996), the most important results of the model are formalised in ten propositions. The next chapter will derive and test hypotheses that are based on only four of these propositions, because the other six are related to issues for which data is unavailable. Hence, only the propositions that are relevant for the empirical analysis of the next chapter are given below.

Proposition I *Initially the number of entrants may rise or decline, but eventually it will decline to zero.*

[60]Here, Klepper follows Flaherty (1980): average cost in period t is a function of only process R&D in period t. However, as Klepper argues: "*If all process improvements were assumed to be costlessly imitated one period after they were introduced, firm differences in average costs would still be a function of only differences in contemporaneous firm spending on process R&D.... This change would not affect the model. Even if cost differences were allowed to cumulate, it would only reinforce the advantages of the largest firms.*"

To prove this proposition, first the first-order conditions for maximising expected current profits are derived by differentiating (1) with respect to rd_{it}, rc_{it}, and Δq_{it}:

$$g'(\mathrm{rd}^*_{it})G = 1 \tag{3.2}$$

$$[Q_{it-1}(Q_t/Q_{t-1}) + \Delta q^*_{it}]l'(\mathrm{rc}^*_{it}) = 1 \tag{3.3}$$

$$m'(\Delta q^*_{it}) = p_t - c + l(\mathrm{rc}^*_{it}), \tag{3.4}$$

where an asterisk denotes optimal values. Since it is assumed that $F > [s_i + g(\mathrm{rd}_{it})]G - \mathrm{rd}_{it}$ for all rd_{it}, a necessary condition for nonnegative expected profits is:

$$p_t - c + l(\mathrm{rc}^*_{it}) > 0 \tag{3.5}$$

This last expression assures that $m'(\Delta q^*_{it}) > 0$ through expression (3.4) and, consequently, that $\Delta q^*_{it} > 0$. Hence all firms that remain in the market in period t increase their market share. Another implication is that the firm's incremental profit earned from product R&D remains constants over time, since rd^*_{it} is defined by (3.2) and is the same for all firms. Let $VE^*_{it} = \Delta q^*_{it} [p_t - c + l(\mathrm{rc}^*_{it})] - \mathrm{rc}^*_{it} - m(\Delta q^*_{it})$ denote the entrant's incremental profit earned from the standard product. Differentiating VE^*_{it} with respect to p_t yields:

$$dVE^*_{it}/dp_t = \partial VE^*_{it}/\partial p_t = \Delta q^*_{it} \tag{3.6}$$

Given that $\Delta q^*_{it} > 0$ and that p_t falls over time, VE^*_{it} must fall over time as well. Thus the marginal entrant must earn greater incremental profit from product innovation over time, which can only be achieved by a higher innovative expertise s_i. Hence, over time the minimum innovative expertise required for entry must rise. As long as this minimum level is lower than or equal to $s_{\max}$, then $E(\Pi^*_{it}) \geqslant 0$ for potential entrants with $s_{\max}$. In that case $E(\Pi^*_{it}) > 0$ for all prior entrants with $s_{\max}$, because they spent more on process R&D and have lower average cost therefore. This can be shown by rewriting (3.3) as $Q_{it}\ l'(\mathrm{rc}^*_{it}) = 1$, and since $Q_{it} > Q_{it-1}$, and $l''(\mathrm{rc}_{it}) < 0$, it follows that $\mathrm{rc}_{it} > \mathrm{rc}_{it-1}$. As a consequence, no incumbent firm with $s_{\max}$ will exit the industry. In fact, they will want to expand their market share, but because this cannot happen indefinitely, some incumbents will have to exit the industry. This requires a decrease in p_t such that $E(\Pi^*_{it}) < 0$ for some incumbents with s_{max}. After that point, potential entrants even with s_{max} will be unable to earn positive profits, and no further entry will occur.

> **Proposition II** *Initially, the number of firms may rise over time, but eventually it will decline steadily.*

In proving proposition I, it was shown that p_t must fall over time by a sufficient amount to cause some firms to exit in every period. Secondly, it was also shown that after some time entry ceases. Hence, eventually the number of firms will decline.

> **Proposition III** *As each firm grows large, eventually the change in its market share will decline over time.*

If Δq_{it} declines in time, $\Delta q_{it}/Q_t$will decline, because Q_t is nondecreasing over time. As expression (3.4) indicates, Δq_{it} is based on $p_t - c + l(rc_{it})$. Since rc_{it} grows in time, $l(rc_{it})$ will grow as well, but eventually it will approach its upper bound. However, p_t falls over time, and therefore $p_t - c + l(rc_{it})$, the profit margin on the standard product, decreases as well. Consequently, Δq_{it} declines in time.

Klepper's next five propositions are all related to issues such as the number of product innovations, and the amount spent on product and process R&D. As mentioned before, the absence of data on these variables impedes the derivation of testable hypotheses.[61] Therefore, these propositions are skipped, which leads us to the fourth proposition considered here (proposition 9 in Klepper's article).

> **Proposition IV** *For each period, firm average cost varies inversely with firm output.*

Klepper (1996, p. 578): "*Since rc_{it} varies directly with Q_{it} and $l'(rc_{it}) > 0$ for all rc_{it}, it follows directly that $c - l(rc_{it})$ varies inversely with Q_{it}.*"

Let us shortly summarise the model. Its main implication is that the industry will eventually evolve to an oligopoly. The earliest successful entrants engage in R&D and start growing. This increases their returns from R&D, because the output in which they can embody their innovations rises.

[61] Audretsch (1987) however found significant differences between industries in different evolutionary stages with respect to R&D and skill intensity "*consistent with the assumptions of the empirical literature.*" (Audretsch, 1987, p. 306).

Initially, i.e., in the expanding stage, firms enter, but at some point the price is driven down to a level at which (profitable) entry is no longer possible. Exit still continues, as for some firms the price falls below their average costs. Hence, the number of firms declines, contributing to a shakeout. Eventually, both industrial sales and the number of firms level off. It is in this consolidating stage that the oligopoly emerges.

3.4.3 Conclusion

By depicting the evolution of the structural and dynamic properties of an industry over its life, theories on product of industry life cycles may explain cross-sectional differences between industries by asserting that the industries are in different stages of their life cycle. This section has provided an overview of these theories, divided into event theories and comparative advantage theories. From the comparative advantage theories, the model by Klepper (1996) was selected to provide the conceptual framework for the empirical analysis of the next chapter. This model provides the most comprehensive theoretical framework as it includes factors as innovation by incumbents, scale economies, and the gradual shift over time from product innovations to process innovations. Furthermore, this models allows for a number of testable hypotheses, as will be shown in the next chapter.

3.5 APPRAISAL

Compared to the equilibrium models described in section 3.2, both the technological regime and the product life cycle approach provide a much more plausible and powerful framework to explain variances in the structures and dynamics of industries. In this section we will first summarise the merits of the two approaches vis-à-vis the mainstream equilibrium models. Next, we will highlight the main difference between them and discuss the extent to which they complement each other. Finally, we will focus on some important issues that are currently still ignored by the technological regime framework and the product life cycle approach.

The merits of the technological regime framework lie predominantly in its ability to identify a number of conditions that together explain why sectoral patterns of innovative activities are different across industries. Bounded rationality, imperfect information and technological uncertainty are important and plausible assumptions of the technological regime framework that are mostly absent in equilibrium models. Given these assumptions, this framework can explain why the creation and diffusion of technological knowledge is an endogenous process that produces winners and losers among the firms involved in research and development activities. As we have argued in section 3.2.4, by assuming that new technological knowledge arrives exogenously, information is perfect and entrepreneurs are perfectly rational, equilibrium models lack the possibility of modelling the competitive process driven by endogenous technological change. Hence, by endogenising technological change and identifying technological innovativeness as a major determinant of the competitive process, the technological regime approach provides a much more comprehensive and plausible framework for analysing the structural and dynamic properties of industries than the equilibrium models.

The strength of the product or industry life cycle approach lies in its ability to endogenise the evolution of an industry over its lifetime. More specifically, it analyses how the terms of (technological) competition change through the formative stages of an industry, and how these changes effect the industry's structural and dynamic properties. To some extent, equilibrium models could be used as well to analyse different equilibrium properties of industries under different competitive and technological[62] conditions. However, these models would not able to explain why the competitive and technological conditions change over time. This is an important omission, because many new industries have shown rather similar evolutionary patterns, suggesting that the competitive process changed in a systematic way as these industries matured. Since models on product life cycles do explain how these changes arise, they provide a much better tool to analyse the evolution of industries than the mainstream equilibrium models.

In conclusion, compared to the mainstream equilibrium models we consider the technological regime and the product life cycle approach much more appropriate and plausible theoretical frameworks for explaining and analysing industrial dynamics. But how do the technological regime framework and the product or life approach compare to each other? In our view, these theoretical bodies are highly complementary. The technological regime framework explains *structural* differences in industrial dynamics *between* industries, whereas the product life cycle approach explains *temporal* differences in industrial dynamics *within* industries. In general, the

[62]In terms of different specifications of the production function.

technological regime framework does not distinguish the various stages through which new industries evolve. Reversely, the product life cycle approach does not distinguish differences in, e.g., cumulativeness conditions between industries. Within the product life cycle approach, the comparative advantage models certainly acknowledge the cumulative nature of the innovative process. But no reference is made to the effect of different degrees of cumulativeness on the evolutionary patterns of the industries.

Since these two approaches are highly complementary, there seems to be a great potential for combining the technological regime framework and the product life cycle approach. To a large extent the deficiencies of the former can be overcome by the latter, and vice versa. The result of such an attempt to combine the two frameworks may be a model that allows for a direct comparison between the life cycle patterns of new industries with different underlying technological regime conditions. However, when the present models of the technological regime and the product life cycle approach were taken as a starting point for such a combined model, a number of important issues would still be ignored.

First of all, models of product life cycles generally focus on the emergence and evolution of only one product and its associated technology. However, in many industries we observe that firms repeatedly introduce or adopt new product technologies that replace the older ones. Second, both these approaches do not explicitly consider differences in the technological properties of the goods produced by the industries. Finally, in models on technological regimes and on industry life cycles the growth of a firm is generally determined by its relative (technological) performance. However, empirical studies on firm growth do not provide much evidence supporting such a relationship. Most of these studies suggest that the size of a firm generally follows a random walk with a declining positive drift.[63]

In conclusion, although the present models of the technological regime and the product life cycle approach still ignore a number of interesting issues, we believe they provide a comprehensive and plausible framework to analyse the structural and dynamic properties of industries. Furthermore, we expect that future efforts to combine them will prove valuable, as these two approaches are highly complementary. A preliminary attempt to combine some parts of the technological regime framework and the product life cycle approach and to overcome their present shortcomings is presented in chapter 5 of this book.

[63]See Geroski (1998).

3.6 CONCLUSIONS

In our search for theoretical contributions that may explain the observed differences across industries in a dynamic and comprehensive way, we have selected two approaches and described them thoroughly in this chapter. In our view, both the technological regime framework as well as the product life cycle approach are highly appropriate conceptual tools to study empirically the differences in cross-section regularities, as they both embody elements such as firm heterogeneity and technological uncertainty that are close to empirical substance. With regard to their potential power to explain differences in structures and dynamics between industries, we have argued the following.

The technological regime framework asserts that patterns of innovative activities are determined by the combination of opportunity, appropriability and cumulativeness conditions, and properties of the technological knowledge base underlying these activities. If entry, survival, growth and exit patterns are related to the rates and forms of organisation of innovations, then the observed inter-sectoral variety in structures and dynamics can be explained by differences in the underlying technological regimes. Alternatively, theories and models on product life cycles explain and depict the evolution of an industry's structural and dynamic properties over its lifetime. Based on this approach, the observed cross-sectional differences can be explained by the different evolutionary stages that industries occupy. To what extent these theories can actually account for the cross-sectional variances will be investigated in the next chapter.

4. TECHNOLOGICAL REGIMES AND INDUSTRY LIFE CYCLES IN DUTCH MANUFACTURING

4.1 INTRODUCTION

The empirical literature in industrial economics has indeed been greatly enriched by the increased availability of longitudinal firm-level databases. As we have shown in chapter 2, the exploration of such databases has resulted in a collection of strong empirical regularities that seem to hold over different countries, as well as over different time periods. One of these regularities concerns the observed variances in the structures and dynamics of industries. However, until now the opportunity to test theories explaining these cross-sectional differences with the available longitudinal firm-level data has hardly been exploited. Using the SN database, the present chapter aims to exploit this opportunity by investigating whether the technological regime framework and the product life cycle approach can explain the observed differences between industries with regard to their structural and dynamic properties.

This chapter is structured as follows. In section 4.2, we will investigate the empirical relationship between industrial dynamics and technological regimes. After deriving a number of hypotheses from the theoretical framework and classifying the industries according to their technological regime, we will test whether the differences suggested by the technological regime framework actually exist. The structure of section 4.3 is rather similar. A number of hypotheses, derived from Klepper (1996), will be tested to see whether industries in different evolutionary stages show different regularities, and whether these regularities are in line with Klepper's model. In section 4.4 we will use regression analyses to investigate the extent to which the selected theoretical approaches simultaneously account for differences between industries, and whether any interaction effects can be observed between them. Section 4.5 concludes this chapter.

4.2 TECHNOLOGICAL REGIMES

As argued in the previous chapter, the main contribution of the technological regime framework is that it suggests an explanation for the observed inter-sectoral variety in the rates and forms of organisation of innovations. Empirical evidence supporting this explanation can be found in Malerba et al. (1995) and Breschi et al. (1996). However, these studies only focused on structural and dynamic differences in the populations of innovative firms in relation to the underlying technological regimes. The question that naturally follows concerns whether a similar relationship exists between the structural and dynamic properties of the full population of firms in an industry and its underlying technological regime.

The aim of the present section is to answer this question. It is structured as follows. First, we will shortly summarise the main characteristics of the technological regime framework, and formulate a number of hypotheses derived from this framework. Second, we will present how we have classified the industries in the SN database according to their technological regime. Finally, we will test the hypotheses with the firm-level data at hand.

4.2.1 The Hypotheses

The previous chapter already discussed in detail the literature on technological regimes. Before formulating the hypotheses, let us briefly summarise the most important issues on the technological regime framework from which the hypotheses will be derived. A technological regime can be defined as a particular combination of opportunity, appropriability, cumulativeness conditions and properties of the knowledge base, common to specific activities of innovation and production and shared by the population of firms undertaking those activities. Opportunity conditions refer to the likelihood of innovating, given a certain research effort. Appropriability conditions reflect the possibilities of protecting innovations from imitation and of appropriating the profits from an innovation. Cumulativeness conditions refer to the extent to which the innovative successes of individual firms are serially correlated. Finally, with regard to the properties of the knowledge base, the following three aspects of knowledge can be distinguished: (1) the level of specificity, (2) the level of tacitness, and (3) the extent to which the knowledge is publicly available.

In the literature on technological regimes, two regimes with opposite underlying conditions are usually distinguished. These metaphorical archetypes correspond to the two rather opposite views on the innovative process that Schumpeter held during his life. The two regimes are labelled Schumpeter Mark I (SM-I) and Schumpeter Mark II (SM-II). Since opportunity conditions do not necessarily differ between the regimes, the differences between them are mainly related to differences in appropriability, cumulativeness conditions and patterns of access to knowledge. A SM-I regime is characterised by low appropriability and cumulativeness conditions, and the knowledge is mainly specific, codified and simple. In a SM-II regime, these three conditions are reversed: appropriability and cumulativeness conditions are high, whereas the knowledge is mainly generic, tacit and complex.

In the presentation of hypotheses that follows, we will base the formulations of the hypotheses on these two archetypes. We will propose one group of industries in which the SM-I regime underlies the innovative and productive activities, and a second group in which the SM-II regime prevails. Based on the differences in the appropriability, cumulativeness and knowledge conditions, we will hypothesise significant differences between the two groups of industries with regard to a number of selected variables. These differences can be classified in two major categories. The first category addresses differences in static measures, such as the firm size distribution, whereas the second category more reflect differences in the dynamic properties of the industry, e.g., the process of entry and exit. We will first derive some hypotheses related to the static inter-industry differences.

As argued in chapter 3, in a typical SM-I industry we can expect the pattern of innovative activities to be characterised by the constant inflow of new entrepreneurs with new ideas, products and processes, continuously challenging the established firms and wiping out their quasi-rents associated with previous innovations. Such an industry is therefore likely to be characterised by the presence of many small firms and low entry barriers (Malerba et al., 1995). In SM-II industries, this situation is reversed: large firms prevail, (technological) entry barriers are high, and the high profit rates of incumbent firms are persistent. Additionally, Audretsch (1997) suggests that the degree of capital intensity also shapes the technological regimes: a routinised technological regime (which resembles a SM-II regime) is more capital intensive than an entrepreneurial technological regime (SM-I regime). All together, these characterisations allow us to derive the following hypotheses:

Hypothesis 1 *The share of small firms is higher in SM-I industries than in SM-II industries.*

Hypothesis 2 *Concentration levels are lower SM-I industries than in SM-II industries.*

Hypothesis 3 *Entry barriers are lower in SM-I industries than in SM-II industries.*

Hypothesis 4 *Capital intensity is lower in SM-I industries than in SM-II industries.*

Hypothesis 5 *In SM-I industries, profit rates are lower than in SM-II industries.*

Hypothesis 6 *In SM-I industries, entrants are more productive than incumbents, whereas in SM-II industries incumbents are more productive than entrants.*

This last hypothesis reflects the relative advantage that entrants should have over incumbents in SM-I industries, and vice versa for SM-II industries. Hence, hypothesis 6 tests for differences between entrants and incumbents *within* each technological regime.

With respect to variances in the dynamic properties of industries, differences in the stability of the firm population can be measured by looking at the amount of turnover (in terms of shifts in market shares) due to entry and exit. Moreover, the turbulence within the group of incumbent firms can be examined here: SM-I industries are expected to show a less stable population of incumbent firms than SM-II industries. This leads to the following hypotheses:

Hypothesis 7 *In SM-I industries, the amount of turnover due to entry and exit is higher than in SM-II industries.*

Hypothesis 8 *The turbulence within the group of incumbent firms is higher in SM-I industries than in SM-II industries.*

Another potential difference between the two groups of industries concerns the share of the entry and exit process in the total growth in productivity. Given the nature of the two regimes, we can expect the contribution this process to productivity growth to be higher in SM-I industries, and the contribution of incumbents to be higher in SM-II industries. Hence, we can derive the following hypothesis:

Hypothesis 9 *The contribution of the entry and exit process to productivity growth is higher in SM-I industries than in SM-II industries, and vice versa for incumbents' contributions.*

In section 4.2.3, these hypotheses will be tested. But first we will explain how we have classified the industries into the groups of SM-I and SM-II regimes.

4.2.2 Classification of Industries

As mentioned in chapter 3, previous empirical research used data on innovation activities performed by the firms to study the technological regime of an industry. Unfortunately, the SN database does not include variables indicating the innovative activities of the firm. We therefore have to rely on the results of previous empirical research to classify our industries into a group of SM-I and a group of SM-II industries.

The classification of industries into technological regimes that we propose is based on the taxonomy presented in Malerba et al. (1995). As mentioned in chapter 3, this classification is based on patent data. By using this patent data, Malerba et al. (1995) constructed a group of empirical proxies of the Schumpeter Mark I (which they call the 'widening' pattern) and Schumpeter Mark II ('deepening') patterns of innovative activities. Next, they grouped the empirical proxies into four types: concentration of innovative activities, size of the innovating firms, change over time in the hierarchy of innovators, and the relevance of new innovators.

After applying these indicators to 33 technological classes, the authors found that two groups of technological classes could be defined. One in which innovative activities are organised according to the widening pattern of innovative activities, and in which innovative activities are organised according to the deepening pattern of innovative activities. It is this division that we have used for classifying the industries in the SN database. Hence, each industry was first allocated to its corresponding technological class and then, based on the taxonomy by Malerba et al. (1995) allocated to the widening pattern (SM-I) or the deepening (SM-II) pattern.

Although this taxonomy does not directly rest on differences with regard to the underlying appropriability, cumulativeness, and knowledge conditions, the applied indicators of innovative activity strongly suggest major differences between the technological classes according to the widening (SM-I) and the deepening (SM-II) pattern. Breschi et al. (1996) found that these patterns are, in turn, related to the conditions of the underlying technological regime. As mentioned in chapter 3, they estimated the impact of a number of technological regime variables on Schumpeterian patterns of innovation (defined by the specific combination of entry, stability and concentration of the innovating firms). Since the authors found considerable evidence for the existence of a relationship between sectoral patterns of technical change and the nature of the underlying technological regime, the classification of industries we propose here sufficiently reflects differences in the underlying appropriability, cumulativeness and knowledge conditions, despite the absence of direct empirical evidence.

Since the 33 technological classes do not always strictly correspond to the industries in our sample, we unfortunately had to exclude those industries for which no evidence existed regarding their underlying technological regime. Hence, only industries that are unambiguously SM-I and SM-II (according to Malerba et al., 1995) are considered in the analysis that follows. Although even for this group of industries there may not be a strict *conceptual* correspondence between a technology class and an industry, we believe this does not cause a major distortion in mapping the technology classes to industries. Furthermore, given the strong similarities across the four European countries investigated by Malerba et al. (1995), we do not expect that a similar exercise based on patent data for the Dutch manufacturing sector will lead to a different taxonomy of technological classes. Finally, given the substantial overlap in the time periods of the dataset by Malerba et al. (1968–1986) and our dataset (1978–1992), we do not believe that the small differences in these time periods will bias the results.

According to the proposed criterion, our total sample of 106 industries is divided in 60 SM-I industries and 18 SM-II industries. The remaining 28 industries are excluded from the analysis in this section. This group of industries mainly involves the manufacturing of food products, which could not be classified in a distinct category in the analysis by Malerba et al. (1995)[64], and the printing industries, which was simply not included in their analysis. The complete list of industries and the technological regimes to which they have been allocated can be found in the Appendix of this chapter, as well as the list of excluded industries.

Before turning to the next section, in which we will test the proposed hypotheses, we would like to repeat our remarks made in chapter 2 with regard to the truncation of the SN database. The point we raised there is that the consequence of this truncation is that our 'entering' and 'exiting'

---

[64]Malerba et al. (1995) put the food and tobacco industry into a group labelled 'other technological classes'.

firms are mainly firms crossing the observation threshold. It might seem awkward then to propose a number of hypotheses related to the process of entry and exit. However, if, for instance, the regime of an industry is favourable to new and small firms, it is likely that it is easier for these firms to grow and pass the observation threshold than in an industry with a regime favourable to large and established firms. To put more strongly, excluding firms with less than twenty employees, even if involving an inevitable censoring of observations, is not likely to distort the entry and exit characteristics associated with the two regimes. Therefore, there is reason to believe that if there are differences between industries with different underlying technological regimes, these differences will be statistically significant, regardless of whether the firm observation threshold is set at twenty, at ten, or at zero employees.

4.2.3 The Results

For every hypothesis, we will calculate the means of the relevant variables for the two groups of industries, and test whether the differences between the means are significantly different between SM-I and SM-II industries, and whether the sign of the difference is consistent with the hypothesis tested. In the tables that follow, the specific variables used for testing the hypothesis can be found in the first column. In the second and third column the means of these variables are listed for both groups of industries, together with their standard errors. The rest of the table shows the output of a two-sided t-test of the null hypothesis that the means of the different groups of industries are the same. Because the computation of the t-statistic depends on whether the variances of the two populations are equal or not, for each of the hypothesis an F-test was performed to test the hypothesis that the population variances are equal. Whenever this hypothesis could be accepted at the 5 percent significance level, the t-statistic (of the hypothesis that the true means of the different groups of industries are the same) was calculated as follows:

$$t = \frac{\bar{x} - \bar{y}}{\sqrt{\frac{(n_x - 1)s_x^2 + (n_y - 1)s_y^2}{n_x + n_y - 2}} \sqrt{\frac{1}{n_x} + \frac{1}{n_y}}},$$

where s is the sample standard deviation, n the number of observations, and x and y are the two groups. This results in a Student's t distribution with

$n-2$ degrees of freedom. When the hypothesis that the population variances are equal could not be accepted at the 5 percent significance level, the t-statistic was calculated as follows:

$$t = \frac{\bar{x} - \bar{y}}{\sqrt{\frac{s_x^2}{n_x} + \frac{s_y^2}{n_y}}}.$$

This again results in a Student's t distribution, with v degrees of freedom, where v is given by Satterthwaite's formula:

$$\frac{\left(s_x^2/n_x + s_y^2/n_y\right)^2}{\frac{\left(s_x^2/n_x\right)^2}{n_x-1} + \frac{\left(s_y^2/n_y\right)^2}{n_y-1}}.$$

In column four the value of the t-statistic is listed. The corresponding p-value can be found in parentheses, indicating the probability of the occurrence of an absolute value larger than t. The lower this p-value, the less likely it is that the true means of the groups of industries under consideration are the same. We reject the hypothesis that the true means are the same when the p-value is less than 0.05.

Hypothesis 1 *The share of small firms is higher in SM-I industries than in SM-II industries.*

The share of small firms, here defined as firms having less than 100 employees, can be measured as the annual mean of their share in the total population of firms, or as the annual mean of their share in total industrial sales. These variables, expressed in percentages, are listed in table 4.1.
As the results in table 4.1 show, there is considerable support for accepting this hypothesis: the share of small firms is significantly higher in SM-I industries.

Hypothesis 2 *Concentration levels are lower in SM-I industries than in SM-II industries.*

To test this hypothesis, we calculated for each industry two concentration measures. The first is the annual mean of the Herfindahl-index, which is the sum of squared market shares of all firms in the industry. The second is the annual mean of the total market share of the largest four firms in the industry (the CR-4 ratio).

As table 4.2 shows, there are significant differences between SM-I and SM-II industries with regard to concentration levels, hence accepting the hypothesis.

Table 4.1 Share of Small Firms

	Mean (standard error)		Group means *T*-test	
Variable	Schumpeter Mark I	Schumpeter Mark II	*T*-statistic	*P*-value
Share of small firms in total population	74.3 (1.96)	52.7 (4.24)	5.06	<0.01
Share of small firms in total sales	42.7 (2.94)	14.5 (2.64)	7.14	<0.01

Hypothesis 3 *Entry barriers are lower in SM-I industries than in SM-II industries.*

Hypothesis 4 *Capital intensity is lower in SM-I industries than in SM-II industries.*

Advertising intensity, capital intensity, and the minimum efficient scale of an industry are often considered as important entry barriers. We cannot test for differences with regard to advertising intensity, since advertising expenditures are not in our data set. Also, there are no data on the capital stock of firms in our set; however we can roughly approximate capital intensity by the average investment margin, i.e., the annual mean of the ratio of investments to sales. Finally, a proxy for the minimum efficient scale is the median plant size of the industry.

As table 4.3 shows, in SM-I industries 'capital intensity' (as approximated here) is not significantly lower than in SM-II industries. However, there is strong evidence that the median size of firms in SM-I industries is lower than in SM-II industries. Hence, the evidence for hypothesis 3 is mixed when *ex ante* measures of entry barriers are applied. By using the proxy for capital intensity as an indicator for entry barriers, hypothesis 4 was automatically tested as well, and rejected. We also measured entry barriers *ex post* by

Table 4.2 Concentration Levels

	Mean (standard error)		Group means *T*-test	
Variable	Schumpeter Mark I	Schumpeter Mark II	*T*-statistic	*P*-value
Herfindahl- index	11.3 (1.29)	18.3 (2.83)	−2.51	0.01
CR-4 ratio	47.1 (2.84)	67.3 (4.03)	−3.59	<0.01

Table 4.3 Entry Barriers

Variable	Mean (standard error) Schumpeter Mark I	Schumpeter Mark II	Group means *T*-test *T*-statistic	*P*-value
Investment margin	6.05 (0.27)	6.63 (0.49)	−1.03	0.31
Median firm size	60.3 (4.01)	121.3 (21.3)	−2.82	0.01
Survival rate	71.0 (2.23)	76.3 (4.12)	−1.15	0.26

calculating for each industry the number of successful entrants (firms not present in 1978, but still present in 1992) relative to the total number of entrants over the whole period. Measured in this way, entry barriers are not significantly higher in SM-II industries than in SM-I industries.

Hypothesis 5 *Profit rates in SM-I industries are lower than in SM-II industries.*

Hypothesis 6 *In SM-I industries, entrants are more productive than incumbents, whereas in SM-II industries incumbents are more productive than entrants.*

For measuring profitability, we use the average profit margin, calculated as the annual mean of the ratio of gross profit to sales. For testing hypothesis 5, the mean of this ratio of all firms in the industry is calculated (expressed in percentages). For hypothesis 6, we use respectively the annual mean of relative labour productivity of all continuing firms, and of all entrants in each industry. A firm's relative labour productivity is equal to its real value added per worker divided by the industry mean. The results are listed in table 4.4.

Hypothesis 5 is accepted: the average profitability in SM-I industries is significantly lower than in SM-II industries. For hypothesis 6, we applied a

Table 4.4 Profitability and Relative Productivity

Variable	Mean (standard error) Schumpeter Mark I	Schumpeter Mark II	Group means *T*-test *T*-statistic	*P*-value
Profit margin	8.52 (0.46)	10.4 (0.67)	−2.09	0.04
Relative productivity of incumbents	1.03 (0.01)	1.01 (0.02)	1.55	0.12
Relative productivity of entrants	1.03 (0.02)	1.07 (0.04)	−0.98	0.33

paired groups t-test. For each industry, the difference between the relative productivity of entrants and incumbents was calculated. Subsequently, for SM-I and SM-II industries the mean of these differences was computed. For both SM-I and SM-II industries, the hypothesis that the true means of these differences are equal to zero was accepted (the t-values are respectively equal to −0.18 and 1.21). Hence, within both groups of industries the relative productivity of incumbents is not significantly different from the relative productivity of entrants.[65] Also note that *between* the regimes, no significant differences exist for the relative productivity of incumbents or entrants.

Hypothesis 7 *In SM-I industries, the amount of turnover due to entry and exit is higher than in SM-II industries.*

There are two ways of measuring the magnitude of the process of entry and exit. The first one simply measures annual firm entry and exit rates, by counting the number of entering and exiting firms and dividing them by the total number of firms active in the industry. Accordingly, the share of sales of entering and exiting firms in total industrial sales can be calculated for each year in order to measure annual sales entry and exit rates. The industrial means of the annual firm entry and exit rates, as well as the industrial means of annual sales entry and exit rates, all in percentages, are listed in table 4.5.

The second way is to measure cumulative entry and exit rates, which reflect the significance of entry and exit in the long run. The cumulative entry rate is the share of entrants in the total number of firms or total sales at the end of a given period, i.e., 1992. The cumulative exit rate is the share of exiting firms in the total number of firms or total sales at the beginning of the period, i.e., 1978. The means of these variables, in percentages, are listed in table 4.6.

As table 4.6 shows, annual and cumulative firm entry and exit rates do not significantly differ between the two groups of industries. However, annual and cumulative sales entry and exit rates are indeed higher in SM-I industries than in SM-II industries. Therefore, when measured in terms of market shares, both for the short and the long run perspective there is considerable evidence that the amount of turnover due to entry and exit rates is higher in SM-I industries than in SM-II industries.[66]

[65]Similar results were obtained for differences between incumbents and entrants with regard to their mean relative profit margins.

[66]Since the number of firms crossing the observation threshold may also depend on the growth or decline of industries, we have tested whether the mean growth rates of total industrial sales were significantly different between the two groups of industries. We have found no evidence for this.

Table 4.5 Annual Entry and Exit Rates

Variable	Mean (standard error)		Group means T-test	
	Schumpeter Mark I	Schumpeter Mark II	T-statistic	P-value
Annual firm entry rate	8.66 (0.50)	8.66 (0.81)	<0.01	0.99
Annual firm exit rate	8.34 (0.44)	7.57 (0.91)	0.82	0.41
Annual sales entry rate	4.84 (0.40)	2.89 (0.40)	3.47	<0.01
Annual sales exit rate	4.97 (0.46)	3.12 (0.69)	2.00	0.05

Table 4.6 Cumulative Entry and Exit Rates

Variable	Mean (standard error)		Group means T-test	
	Schumpeter Mark I	Schumpeter Mark II	T-statistic	P-value
Cumulative firm entry rate	45.9 (2.21)	39.7 (3.66)	1.37	0.17
Cumulative firm exit rate	37.4 (1.76)	32.7 (3.96)	1.22	0.23
Cumulative sales entry rate	34.6 (2.55)	18.0 (3.11)	3.34	<0.01
Cumulative sales exit rate	34.3 (2.72)	16.3 (3.23)	3.41	<0.01

Hypothesis 8 *The turbulence within the group of incumbent firms is higher in SM-I industries than in SM-II industries.*

Two measures are proposed here. The first one measures the short run turbulence within the group of incumbent firms. For each year and for each continuing firm (i.e., firms present in 1978 and 1992), the absolute change in market share is calculated. Next, for each industry the sum of the absolute changes in market shares is computed for each year. Finally, the annual mean of the sum of changes is calculated for each industry. The second measure measures the long run turbulence. For each continuing firm, the absolute change in market share between 1978 and 1992 is calculated. The long run turbulence of an industry is then equal to the sum of absolute market share changes of the continuing firms within that industry. The results are listed in table 4.7.

Only when measured in the short run, the amount of turnover is significantly higher in SM-I industries than in SM-II industries. The long run measures of turbulence are not significantly different between the two groups of industries.

Table 4.7 Turbulence Among Incumbents

Variable	Mean (standard error)		Group means T-test	
	Schumpeter Mark I	Schumpeter Mark II	T-statistic	P-value
Short-run turbulence	14.4 (0.87)	11.1 (0.50)	3.34	<0.01
Long-run turbulence	26.6 (1.35)	28.9 (2.25)	-0.87	0.39

Hypothesis 9 *The contribution of the entry and exit process to productivity growth is higher in SM-I industries than in SM-II industries, and vice versa for incumbents' contributions.*

For calculating the contribution of the entry and exit process to the total productivity growth of an industry, we applied the same decomposition method[67] as the one used for decomposing productivity growth at the manufacturing level in section 2.3.2 of chapter 2. The contribution of the entry and exit process is then calculated as the sum of the entry and exit effect. For each industry with positive productivity growth[68], we have calculated these numbers and divided them by the total productivity growth of each industry in order to get their shares in total productivity growth. In table 4.8 the means of these shares are listed for both groups of industries.

If we compare the difference in the contribution of the entry and exit process between the two regimes, we find strong support for hypothesis 9. The contribution to total productivity growth of entrants is significantly higher in SM-I industries than in SM-II industries. The results also indicate that *within* the group of SM-I industries, the contributions of entry and exit on the one hand, and the contribution of the group of continuing firms on the other[69], are not significantly different. This is confirmed by a paired groups t-test, similar to the one applied at testing hypothesis 6 (t-value equal to 0.50). However, within the group of SM-II industries, the contribution by entering and exiting firms is much lower than the contribution by the group of incumbent firms (t-value equal to −4.13).

[67]Originally taken from Haltiwanger (1997).

[68]This restriction ruled out 13 SM-I industries and 1 SM-II industry.

[69]The contribution of the group of incumbent firms to total productivity growth is simply 100 percent minus the contribution of the entry and exit process.

Table 4.8 Contribution of Entry and Exit to Productivity Growth

	Mean (standard error)		Group means *T*-test	
Variable	Schumpeter Mark I	Schumpeter Mark II	*T*-statistic	*P*-value
Share in productivity growth of the entry and exit process	51.0 (11.7)	14.8 (8.53)	2.50	0.02

In conclusion, most of the hypotheses of this section are accepted. The two groups of industries show significantly different group means with regard to most of the selected variables. However, one objection to our analysis might be that these results are merely the outcome of the aggregation of heterogeneous industries. Depending on the distributions of the industry-level variables over the industries, a considerable number of other (random) partitions of the industries in the sample may have led to similar conclusions regarding the significance of the differences in group means. In order to investigate whether this objection is right, one could calculate for each variable the probability of finding (by chance alone) a partition of the 78 industries into a group of 60 and 18 industries yielding significantly different group means as well. One way to do this would be to consider all possible partitions[70], and count how many of these exhibit significant differences in group means. If this number was high relative to the total number of possible partitions, the original partition based on the taxonomy by Malerba et al. (1995) may not be so 'unique'. In that case many other partitions would lead to significantly different group means as well, rendering the distinguishing power of the original partition moot.

Given the extremely large total number of possible partitions, performing the exercise mentioned above obviously requires too much computation time. Instead, we have approximated the true proportion (of the total number of possible partitions) that yields significantly different group means by the following procedure. In total 25,000 times, we have randomly divided the 78 industries into a group of 60 and a group of 18 industries, and counted the number of times the means of the industry-level variables[71] were sig-

[70]In total, there are 2.12×10^{17} possible partitions $\left(\frac{78!}{60! \times 18!}\right)$.
[71]We have omitted the variables that tested for differences *within* the technological regimes, as these variables are not included in hypotheses that test for differences *between* the SM-I and SM-II groups.

Table 4.9 Analysis of Random Grouping Procedure

Variable	Significantly negative	Significantly positive	Variable	Significantly negative	Significantly positive
Share of small firms in total population	3.39%	2.98%*	Annual sales entry rate	3.23%	3.80%*
Share of small firms in total sales	2.99%	2.74%*	Annual sales exit rate	3.63%	4.75%*
Herfindahl-index	3.38%*	5.23%	Cumulative firm entry rate	2.63%	2.81%*
CR-4 ratio	2.75%*	2.70%	Cumulative firm exit rate	2.58%	3.11%*
Investment margin	2.82%*	3.50%	Cumulative sales entry rate	2.81%	2.74%*
Median firm size	3.98%*	8.35%	Cumulative sales exit rate	3.01%	2.66%*
Survival rate	6.68%	3.12%*	Short-run turbulence	3.83%	6.60%*
Profit margin	2.84%*	2.74%	Long-run turbulence	2.99%	2.83%*
Annual firm entry rate	3.32%	3.51%*	Share in productivity growth of the entry and exit process		
Annual firm exit rate	3.08%	3.56%*		4.81%	7.27%*

Note: Assuming that the SM-I regime underlies the large group, an asterisk indicates the sign for each variable that would be consistent with the technological regime framework.

nificantly different between the two subgroups. For each variable, table 4.9 lists the percentage of cases in which the mean of the 60 industries group is significantly lower than the mean of the 18 industries group, and the percentage of cases in which the opposite holds.[72] Hence, the sum of these percentages approximates the probability of finding by chance alone a partition yielding significantly different group means (regardless of their signs).

As the results in table 4.9 show, the proportion of cases in which the random grouping procedure leads to significantly different group means consistent with the technological regime framework is quite low. And given that the variables are not perfectly correlated, we can expect the number of cases in which the means of a selection of these variables are simultaneously

[72]Based on the *t*-test described in section 4.2.3.

different *and* consistent with the technological regime framework to be even much lower.[73] In conclusion, the results of this random grouping procedure provide additional evidence that the criterion based on the taxonomy by Malerba et al. (1995) constitutes a significant determinant of structural and dynamic differences between industries.

4.2.4 Conclusion

The main conclusion from the analysis of this section is that the results indeed suggest that the structural and dynamic properties of the industries are strongly related to their underlying technological regimes. Obviously, the two groups of industries show considerable differences with regard to most of the selected variables, which in all cases have the right sign, and thus corroborate the hypotheses. For only a few variables the differences are not significant. Let us shortly summarise the results.

With regard to firm size distribution, we have found considerable evidence for the hypothesis that the share of small firms is higher in SM-I industries than in SM-II industries. Furthermore, significant differences were found for the concentration levels of industrial sales. The two groups of industries also showed a considerable difference with respect to the median firm size. However, using a proxy for capital intensity as an alternative measure for entry barriers did not provide any evidence that entry barriers are higher in SM-II industries than in SM-I industries. Overall profitability is significantly higher in SM-II industries, however incumbent firms are not more productive (or profitable) than entrants in SM-II industries, and vice versa in SM-I industries.

Regarding the dynamic measures, we have found that both short run as long run indicators show that the amount of market share turnover due to entry and exit is higher in SM-I industries than in SM-II industries. Interestingly, however, this goes together with the evidence that firm entry and exit rates did not differ significantly between the two groups of industries. The measures of turbulence within the group of incumbent firms showed some mixed results. This variable was only significantly different when measured as year to year changes. If the full period was taken into account, the turbulence within the group of incumbents is approximately the same for SM-I and SM-II industries. Finally, the contribution of the entry and exit

[73]In fact, none of the 25,000 random partitions exhibited a sequence of group means test results for the 19 variables identical to the sequence derived from the original analysis of this section.

process to total productivity growth is higher in SM-I industries than in SM-II industries.

These results show that technological regimes to a considerable extent explain differences between industries with regard to their structural and dynamic properties. In our view, the results of this analysis are strong enough to support to "... *interpretations leaning toward an evolutionary approach* [which] *emphasize sector/technology-specific patterns and invariances in the way agents learn ... as determinants of both structures and dynamics.*" (Dosi et al., 1997, p. 19). In the next section we will investigate to what extent differences between industries can be explained by the different evolutionary stages they occupy.

4.3 INDUSTRY LIFE CYCLES

This section attempts to interpret the observed differences between industries with regard to their structural and dynamic properties by relating them with theoretical conceptualisations regarding industry life cycles. By asserting that industries subsequently go through their stages of birth, growth, maturity and decline, theories on product or industrial life cycles could explain the observed differences in the following way. If each of these evolutionary stages of an industry has its own structural and dynamic properties, then cross-sectional differences between industries may stem from the different evolutionary stages those industries are in. This section aims to find empirical evidence for this explanation by posing the following question: can differences between industries be explained by the different evolutionary stages they occupy?

In order to answer this question, we will derive and test a number of hypotheses based upon Klepper's model, as presented in chapter 3. The structure of this section will be as follows. First, we will briefly explain how we will employ Klepper's model in the empirical analysis of this section, and present the hypotheses derived from this model. Next, we will present how we have assigned the evolutionary stages to the industries in the SN database, and discuss some restrictions the data may impose on testing the hypotheses. Finally, we will test the hypotheses and discuss the results.

4.3.1 The Hypotheses

As mentioned in chapter 3, the most important feature of Klepper's model is the dynamic increasing returns from technological change. The model emphasises differences in firm innovative capabilities, and the importance of firm size in appropriating the returns from innovation to explain regularities concerning how entry, exit, market structure, and innovation vary from the birth of technologically progressive industries through maturity. The main implication of this model is that the industry will eventually evolve to an oligopoly. The earliest successful entrants engage in R&D and start growing. This increases their returns from R&D, because the output in which they can embody their innovations rises. Initially, i.e., in the expanding stage, firms enter, but at some point the price is driven down to a level at which (profitable) entry is no longer possible. Exit still continues, as for some firms the price falls below their average costs. Hence, the number of firms declines, contributing to a shakeout. Eventually, both industrial sales and the number of firms level off. The oligopoly emerges in this consolidating stage.

Unfortunately, the time span of the SN manufacturing database does not allow for studying the full evolution of the industries in the sample. In fact, it is likely that most of these industries have already reached maturity. Furthermore, the level of aggregation of the sampled industries may in some cases be too high to investigate the impact of life cycles of individual products. Therefore, we do not aim to directly test the model. Rather, we try to investigate whether basically mature industries exhibit distinct evolutionary stages as well, and whether dynamic increasing returns to technological change underlie the emergence of these stages. For instance, the obsolescence of the prevailing design of a good, or a radical change in the technology used to produce it may initiate a new evolutionary pattern in a mature industry, resembling the life cycle of a new industry. Hence, we will employ Klepper's model as a conceptual tool to answer the following questions: (1) Do industries in different stages show different regularities? (2) Are these observed regularities in line with the regularities predicted by Klepper's model? (3) Can the observed differences be explained by dynamic increasing returns from technological change?

We will try to answer these questions by testing a number of hypotheses derived from some selected propositions in Klepper (1996). In chapter 3 these propositions and their formal proofs were already presented, however, in what follows we will repeat the selected propositions and present the hypotheses derived from these propositions. We will hypothesise differences between a group of expanding industries and a group of industries experiencing a shakeout. In section 4.3.2 we will explain how we have classified the industries into these groups.

Proposition I *Initially the number of entrants may rise or decline, but eventually it will decline to zero.*

As we argued in section 4.2, differences between entry and exit rates with the SN database are in fact differences in the extent to which firms cross the observation threshold. Therefore, we cannot test Klepper's first proposition directly. We can, however, investigate differences between the stages with regard to the proportion of firms entering the sizeclass of more than twenty employees. If the advantage of incumbents over entrants grows over time, it should become more difficult for small firms to (1) grow and pass the observation threshold, (2) to stay above the threshold, and (3) to obtain a significant market share in the industry. This leads to the first three hypotheses.

Hypothesis I.1 *The entry rates in expanding industries are higher than in industries experiencing a shakeout.*

Hypothesis I.2 *Firms entering expanding industries have higher survival rates than firms entering shakeout industries.*

Hypothesis I.3 *The collective share of all entrants at the end of the observation period is higher in expanding industries than in industries experiencing a shakeout.*

Klepper's assumption of profit maximisation assures that all firms that remain in the market increase their market share.[74] Therefore:

Hypothesis I.4 *Incumbents have a higher market share in 1992 than in 1978.*

Since the model predicts that over time incumbents achieve an increasing cost advantage over entrants, we can derive a fifth hypothesis:

Hypothesis I.5 *Incumbents are more productive than entrants, and the productivity gap between continuing firms and entrants is higher in shakeout industries than in expanding industries.*

To summarise these five hypotheses: testing I.1 to I.3 should reveal the growing advantage of incumbents over entrants, whereas testing I.5 should indicate whether this growing advantage is caused by technological

---

[74]In terms of the model: $\Delta q_{it}^{*} > 0$; see section 3.4.2.

superiority in production. Hypothesis I.4 follows directly from the model's assumption of profit maximisation.

Proposition II *Initially, the number of firms may rise over time, but eventually it will decline steadily.*

Since we will use the change in the number of firms for classifying the industries into the stages of the life cycle, this proposition cannot be tested. However, Klepper's following intuitive proof of this proposition allows for another hypothesis.

Klepper (1996, p. 573): "*Intuitively, over time price falls, the more innovative incumbents expand, and the less innovative incumbents exit and are replaced by more innovative, smaller entrants. This can result in a rise in the number of producers. However, as incumbents continue to grow their advantage eventually becomes insurmountable and entry ceases. Exit continues, though, as the largest firms with the greatest innovative expertise expand their market shares and push the less fit firms out of the market. Consequently, eventually the number of firms declines over time.*"

Hypothesis II.1 *In expanding industries, a certain number of incumbents is replaced by a larger number of smaller entrants.*[75] *In industries experiencing a shakeout, the largest firms expand their market shares at the expense of smaller firms.*

Proposition III *As each firm grows large, eventually the change in its market share will decline over time.*

The prediction that the continuing firm's growth in market share declines over time can easily be tested with the available data. However, testing whether the profit margin on the standard product decreases is more difficult. In the model, the profit margin is defined as price minus average cost of production. The latter are negatively dependent on the amount spent on process R&D. Since there is no data on any of these variables, the profit margin (total profits before taxes over total sales) is taken as a proxy for the profit rate. This leads to the following hypotheses.

[75]The emphasis here is obviously not on differences in sizes between entrants and exiting firms, simply because the nature of the data does not allow for reasonably testing of this hypothesis. What will actually be tested here is whether some incumbent firms exit the expanding industry, whereas at the same time new firms enter.

Hypothesis III.1 *The growth of continuing firms' market shares is positively related to their profit rates.*

Hypothesis III.2 *Profit rates are higher in expanding industries than in industries experiencing a shakeout.*

Hypothesis III.3 *The growth of market shares of continuing firms is higher in expanding industries than in industries experiencing a shake-out.*

Proposition IV *For each period, firm average cost varies inversely with firm output.*

This proposition implies that in each period small firms are less productive than large firms, basically because of scale economies in process R&D. Hence, the following hypothesis can be derived.

Hypothesis IV.1 *For each period, large firms are more productive than small firms.*

In section 4.3.3 we will test these hypotheses and discuss the results. But first we will explain how we have classified the industries into their evolutionary stages.

4.3.2 Classification of Industries

Given the characteristics of the SN dataset, a number of problems arise. Obviously, the major problem is that the entire evolutionary history of each industry cannot be observed, i.e., the origin of the industries occur before the first year of observation. Therefore, it cannot be tested whether the industries evolve according to the pattern predicted by Klepper's model. Instead, we will attempt to locate the industries in the sample in terms of the evolutionary stage that they occupy in Klepper's framework.[76] This is facilitated by the relatively long period over which the firms are observed in the dataset.

Ideally, some independent proxies, such as the growth rate of patents, would be used to determine the evolutionary stage the industry is in. Since these, or any other data that would allow us to exogenously classify the industries, are not available, we will have to base our classification on the revealed patterns of some variables denoting the evolution of the industries

[76]This methodology is very similar to Audretsch (1987).

in the sample. Obviously, such an endogenous classification will, to some extent, limit our empirical analysis. However, we can still investigate whether the industry patterns, denoted by variables other than the ones used for classifying the industries, are consistent with the regularities predicted by Klepper's model. If we would find that the patterns are indeed consistent, such results would provide evidence, albeit circumstantial, that differences in regularities between industries in different stages can be explained by dynamic increasing returns to technological change.

We propose a combination of the following two variables to determine in which stage an industry is: the growth rate of the total number of firms and the growth rate of the sum of their total industrial real sales, both measured over the full period (i.e., 1978–1992). A standard *t*-test was applied to determine whether each of these growth rates was significantly different from zero. Table 4.10 explains how each industry is classified into the stage of its evolution. In parentheses the number of industries in each category is listed.

Given this classification, we have 51 industries in the expansion phase, 28 industries experiencing a shakeout, 6 industries are consolidating, 17 industries are contracting, and 4 industries are in the 'other' class. This last group and the group of consolidating industries will be left out of the analysis, because of the small sample sizes. The high standard errors of the means of most of the variables impede the detection of significant differences. Hence, the focus will only be on expanding industries, industries experiencing a shakeout, and contracting industries. In the Appendix of this chapter all industries are listed and grouped according to their evolutionary stage.

A further limitation of the data to the empirical analysis of this section is the definition of an industry. In Klepper (1996), an industry consists mainly

Table 4.10 Classification of the Industries

		Growth rate of the total number of firms		
		Significantly positive	Approximately zero	Significantly negative
Growth rate of total industrial sales	Significantly positive	Expanding (36)	Expanding (15)	Shakeout (16)
	Approximately zero	Other (3)	Consolidating (6)	Shakeout (12)
	Significantly negative	Other (1)	Contracting (4)	Contracting (13)

of firms producing the same standard product and of some firms temporarily producing a distinctive variant of the standard product. Although the industry classification of the Statistics Netherlands dataset is designed to obtain high levels of homogeneity with respect to the products and the technologies used within the industries, there might still be a number of industries left that are less homogeneous than the industries in Klepper's model. Since the identification of those industries is to some extent an arbitrary exercise, and in order to keep the sample large, no attempt has been made to identify or exclude those industries. For the same reasons, no attempt is made to exclude industries that are not technologically progressive.[77]

Since the industries might show different structural properties, especially with respect to (potential) entry barriers, dividing the industries in three subgroups could create some bias. For instance, if by coincidence the group of expanding industries contains a large number of industries with high entry barriers, then the results of testing a hypothesis for differences between the group of industries with respect to entry rates will be unreliable. Therefore, we have compared the three groups with respect to the means of capital intensity, approximated by the ratio of investments to sales, and median firm size. Because the means of these variables do not differ significantly between the three groups, there is less reason to believe that the proposed division leads to a bias with respect to different structural properties of the groups of industries.

4.3.3 The Hypotheses Tested

In this section, the ten hypotheses will be tested. For each hypothesis, the specific variables used for testing will be defined. Subsequently, the results of the test will be reported, together with a short discussion on their consistency with the model. Although Klepper's model does not deal with the evolution of industries after consolidation (the total quantity demanded expands over time), it might be interesting to see to what extent the group of contracting industries differs from the other groups. Therefore, for each hypothesis the values of the specific variables for the contracting industries will be listed as well.

[77]I.e., we have not attempted to identify and exclude industries that do not manufacture "*products with rich opportunities for both product and process innovation*" (Klepper, 1996).

Table 4.11 Firm Entry Rates

		T-statistic (*p*-value) of a *T*-test for mean differences	
Industry stage	Mean (std. err.)	vs. Shakeout-industries	vs. Contracting industries
Expansion	9.81 (0.72)	4.06 (<0.01)	1.15 (0.25)
Shakeout	6.48 (0.39)		−1.46 (0.16)
Contraction	8.20 (1.10)		

In each of the following tables that show the results of testing for differences between the groups of industries, the mean and its standard error of the specific variable will be listed in the second column. The rest of the table shows the output of a *t*-test of the null hypothesis that the means of the different groups of industries are the same.[78] In the third and fourth column the values of the *t*-statistic will be listed. The corresponding *p*-values can be found in parentheses, indicating the probability of the occurrence of a value larger than *t*. We reject the hypothesis that the true means are the same when the *p*-value is less than 0.05.

Hypothesis I.1 *The entry rates in expanding industries are higher than in industries experiencing a shakeout.*

For each industry, the mean of annual gross entry rates, defined as the number of entrants in a certain year divided by the total number of firms present at the beginning of the year, is calculated.

As the *t*-statistic indicates, the gross entry rates are indeed significantly higher in expanding industries than in industries experiencing a shakeout of firms. Hence, this hypothesis is accepted. Contracting industries show intermediate values for this variable, but with the highest standard error of the mean. The mean of the gross entry rates of contracting industries is not significantly different from the means of the groups of expanding and shakeout industries.

To some extent the acceptance of this hypothesis may seem too obvious: one could argue that it is not surprising to see higher entry rates in industries in which the total number of firms grows, as compared to industries in which this number declines. However, the classification of industries is based on *net* entry, whereas this hypothesis tests for differences with regard to *gross* entry. Hence, in principle shakeout industries may even show higher entry

[78]For the details of this test, see section 4.2.3 of this chapter.

Table 4.12 Firm Exit Rates

Industry stage	Mean (std. err.)	*T*-statistic (*p*-value) of a *T*-test for mean differences vs. Shakeout-industries	vs. Contracting industries
Expansion	7.87 (0.81)	−0.01 (0.99)	−1.83 (0.07)
Shakeout	7.88 (0.47)		−2.99 (<0.01)
Contraction	10.6 (0.90)		

rates than expanding industries. In that case of course the exit rates of shakeout industries would have to significantly exceed the exit rates of expanding industries as well.

Table 4.12, in which we have displayed the means of annual firm exit rates, shows that this is not the case after all. On average, firm exit rates are not significantly higher in industries experiencing a shakeout than the exit rates for the group of expanding industries. But the interesting conclusion is then still that, exactly as in Klepper's model, a shakeout of firms is not caused by an increase in exit rates, but by a significant reduction in the number of firms entering the industry.

Another argument against the acceptance of this hypothesis might be that expanding industries have higher growth rates (in terms of sales) then industries experiencing a shakeout, and that therefore entry rates are higher as well. However, if this were true, contracting industries would have to show the lowest entry rates. Since this is not the case, the lowest entry rates of shakeout industries may be interpreted as circumstantial evidence for an advantage of incumbents over entrants in this group of industries.

Finally, one could argue that the emergence of niches, for instance through specialisation, has been much higher in expanding industries. However, this is not confirmed by the latest standard classification of industries, introduced by Statistics Netherlands in 1993. When the total sales of a new product, initially assigned to existing industries, become sufficiently high, Statistics Netherlands introduces a new 4-digit industry. If this has happened relatively often in expanding industries, it could explain the higher entry and survival rates. This is not the case, however. About one fifth of the group of expanding industries has been split up in two or more new 4-digit industries[79], equal to the proportion of the group of shakeout industries that have been subdivided in 1993.

[79]As compared to the 1974 classification used in the present analysis.

Table 4.13 Survival Rates

		T-statistic (p-value) of a T-test for mean differences	
Industry stage	Mean (std Err.)	vs. Shakeout-industries	vs. Contracting industries
Expansion	77.1 (1.82)	3.46 (<0.01)	2.24 (0.04)
Shakeout	61.1 (4.28)		−0.76 (0.45)
Contraction	66.1 (4.60)		

Hypothesis I.2 *Firms entering expanding industries have higher survival rates than firms entering shakeout industries.*

The variable used here is the industrial mean of the survival rates. It is calculated (for each industry) as the number of successful entrants (firms not present in 1978, but still present in 1992) relative to the total number of entrants over the whole period.

Again, this hypothesis is accepted. Survival rates are on average significantly higher in expanding industries than in industries experiencing a shakeout. Contracting industries show a value for this variable similar to the one for the group of shakeout industries.

Hypothesis I.3 *The collective share of all entrants at the end of the observation period is higher in expanding industries than in industries experiencing a shakeout.*

For testing this hypothesis the cumulative sales entry rate is calculated for each industry. It is the mean of the total market share in 1992 of all (surviving) firms that have entered the dataset between 1978 and 1992.

Just like the two previous hypotheses, hypothesis I.3 is accepted as well. In expanding industries, on average almost one third of the total industrial

Table 4.14 Cumulative Sales Entry Rates

		T-statistic (p-value) of a T-test for mean differences	
Industry stage	Mean (std err.)	vs. Shakeout-industries	vs. Contracting industries
Expansion	34.8 (2.93)	3.02 (<0.01)	2.13 (0.04)
Shakeout	21.2 (2.88)		−0.43 (0.67)
Contraction	23.2 (3.48)		

sales in 1992 is captured by firms that entered the dataset between 1978 and 1992, which is significantly higher than in shakeout industries. For contracting industries, the mean is about the same as for industries in the shakeout stage, and it is significantly lower than for the group of expanding industries.

The results of testing the first three hypotheses indeed seem to indicate a growing advantage of incumbents over entrants. The evidence suggests that it is more difficult for small firms to grow and pass the observation threshold, to stay above the threshold, and to obtain a significant market share in the industry when expanding and shakeout industries are compared. Contracting industries show values that are not significantly different from the values found for the group of shakeout industries.

Hypothesis I.4 *Incumbents have a higher market share in 1992 than in 1978.*

For each continuing firm present in expanding or shakeout industries, the difference in market shares between 1992 and 1978 is calculated. The mean over all incumbent firms of this difference is −0.09 percent (t-value is equal to −1.44). Of all 1962 incumbents in this sample, 907 firms had a higher market share in 1992. On average, they increased their market share by 1.2 percent. The other 1055 firms lost on average 1.2 percent.

Another way of testing this hypothesis is by measuring difference between the collective share of all incumbents in 1992 and 1978. The mean over all industries of this difference is 2.3 percent, but with a t-value of −1.14 it is not significantly different from zero. Hence, there is no support for this hypothesis.

Hypothesis I.5 *Incumbents are more productive than entrants, and the productivity gap between continuing firms and entrants is higher in shakeout industries than in expanding industries.*

The relative productivity of entrants is the mean value of productivity of entrants divided by the mean for all firms (including exiting firms) in the industry.[80] The same procedure is applied to the calculation of the relative

[80]Here, unweighted means were used: the mean value of, for instance, entrants' productivity is the unweighted mean of all observations available for successful entrants within an industry. When weighted means are used, approximately the same results are observed. Still, unweighted means are preferred as they mitigate the effect of the appearance of very large firms in an industry, which in most cases are no entrants in this dataset.

Table 4.15 Relative Productivity Incumbents and Entrants

Industry stage	Relative productivity of incumbents	Relative productivity of entrants	Paired groups *T*-test	
			T-statistic	*P*-value
Expansion	1.02 (0.01)	1.03 (0.01)	−0.34	0.74
Shakeout	1.04 (0.01)	0.99 (0.05)	1.02	0.32
Contraction	1.05 (0.02)	1.04 (0.06)	0.11	0.91

productivity of incumbent firms. In table 4.15 the mean of these variables are presented for the three stages, together with their standard errors.

This hypothesis is rejected: for each group of industries, incumbent firms are on average not significantly more productive than entrants. This is confirmed by a paired groups *t*-test. For each of the three groups of industries, the hypotheses that the true means of the differences between the relative productivity of incumbents and entrants are equal to zero could not be rejected on the basis of the *t*-values.[81] This result is important, because it possibly rejects the hypothesis that the growing advantage of incumbents over entrants is in general due to increasing returns to technological change.

> **Hypothesis II.1** *In expanding industries, a certain number of incumbents is replaced by a larger number of smaller entrants. In industries experiencing a shakeout, the largest firms expand their market shares at the expense of smaller firms.*

Four measures are used. The first one is the number of entrants over the number of exiters per industry. Entrants are here defined as firms not present in 1978, but present in 1992. Exiters are firms present in 1978, but absent in 1992. As mentioned before, the observation threshold of the data used here impedes reasonable size comparisons between entering and exiting firms. But perhaps it is interesting then to compare the ultimate size of entrants with the initial size of exiters. The former is the industrial mean of the 1992 size of entrants (in terms of the number of employees); the latter is the industrial mean of the 1978 size of exiters. For testing the second part of hypothesis II.1, the absolute change in the Herfindahl index is used. If large

[81] We have also tested for differences in relative profitability between incumbents and entrants. For this variable, no significant differences were found as well.

Table 4.16 Entrants Versus Exiters and Changes in Concentration Levels

Industry stage	Number of entrants over the number of exiting firms	Ultimate size of entrants	Initial size of exiters	Absolute change in the Herfindahl-index
Expansion	2.80 (0.36)	68.0 (4.27)	92.2 (13.4)	−2.39 (1.03)
Shakeout	0.74 (0.09)	100.8 (25.3)	101.7 (17.1)	1.42 (1.05)
Contraction	0.83 (0.23)	64.2 (6.16)	161.7 (29.0)	1.32 (1.79)

firms indeed expand their market shares at the expense of smaller firms, this index should increase.

This hypothesis is not confirmed by the data. Naturally, in expanding industries a number of exiting firms is replaced by a higher number of entrants. However, the 1992 size of these entrants is not significantly[82] lower than the 1978 size of exiters. With regard to the second part of hypothesis II.1, no evidence is found. On average, the Herfindahl index does not increase significantly in shakeout industries. An interesting observation is that in contracting industries, the initial size of exiters is significantly larger than the ultimate size of entrants.

Hypothesis III.1 *The growth of continuing firms' market shares is positively related to their profit rates.*

For each continuing firm present in the group of expanding and shakeout industries, the change in market share between 1978 and 1992 was calculated. Then, again for each firm, the average profit rate was calculated by the sum of real profits (between 1978 and 1992) divided by the sum of real sales over the same period. This average profit rate of each incumbent was then divided by the average industrial profit rate for continuing firms to calculate the relative profitability of each continuing firm. To test hypothesis III.1, the correlation coefficient between the change in market share and the log of relative profitability was calculated. Although the coefficient is significantly positive at the 1 percent level, its value of 0.11 is not high. But despite this low value, the hypothesis is accepted.

Hypothesis III.2 *Profit rates are higher in expanding industries than in industries experiencing a shakeout.*

[82]As shown by a paired groups *t*-test, similar to the test applied to hypothesis I.5.

Table 4.17 Profit Margins

		T-statistic (*p*-value) of a *T*-test for mean differences	
Industry stage	Mean (std. err.)	vs. Shakeout-industries	vs. Contracting industries
Expansion	9.48 (0.42)	0.51 (0.61)	2.42 (0.02)
Shakeout	9.10 (0.67)		1.60 (0.12)
Contraction	7.35 (0.88)		

The profit rates are approximated by the profit margin: the total profits (before taxes) over total sales. Then for each industry the mean of these profit margins is calculated.

Because the differences are insignificant between expansion and shakeout, hypothesis III.2 is not accepted. Contracting industries exhibit the lowest profit rates, which are significantly lower than the profit rates of expanding industries.

Hypothesis III.3 *The growth of market shares of continuing firms is higher in expanding industries than in industries experiencing a shake-out.*

For testing this hypothesis, the same variables defined at hypothesis I.4, both expressed in percentages, are used.

The tables show exactly the opposite of what is stated in hypothesis III.3. The mean of the growth of market shares of incumbents is significantly lower (it is even negative) in expanding industries than in shakeout industries. The highest values for these variables are found in contracting industries. These results reject hypothesis III.3, however they do not necessarily reject the hypothesis that incumbents have a growing advantage over entrants.

Table 4.18 Firm-level Differences in Market Shares of Continuing Firms

		T-statistic (*p*-value) of a *T*-test for mean differences	
Industry stage	Mean (std. err.)	vs. Shakeout-industries	vs. Contracting industries
Expansion	−0.44 (0.32)	−2.48 (0.02)	−3.28 (<0.01)
Shakeout	0.87 (0.40)		−1.98 (0.06)
Contraction	3.00 (1.00)		

Table 4.19 Differences in the Collective Market Shares of Continuing Firms

		T-statistic (*p*-value) of a *T*-test for mean differences	
Industry stage	Mean (std. err.)	vs. Shakeout-industries	vs. Contracting industries
Expansion	−9.20 (2.12)	−5.43 (<0.01)	−8.42 (<0.01)
Shakeout	10.3 (2.92)		−3.63 (<0.01)
Contraction	29.5 (4.78)		

Table 4.20 Growth of Industrial Sales vs. Incumbents' Sales

Industry stage	Growth rate of total industrial sales	Growth rate of the collective sales of continuing firms
Expansion	0.91 (0.09)	0.52 (0.05)
Shakeout	0.21 (0.04)	0.44 (0.07)
Contraction	−0.41 (0.04)	0.20 (0.23)

To show this more clearly, two variables are calculated for each industry and listed below in table 4.20. The first one is the mean of the growth rates of real total industrial output. The second variable is the growth rate (between 1978 and 1992) of the real sales of all continuing firms together. Between parentheses the standard errors of the means are listed.

In expanding industries, total industrial output grew by 91 percent on average. The collective sales of continuing firms, however, grew only by 52 percent. Apparently, continuing firms do benefit from the growth of the industry, but their collective share grows at a slower rate than total industrial output. But in contracting industries, the mean growth rate of total industrial output is minus 41 percent, whereas continuing firms together increase their sales by 20 percent on average (the standard error is quite high, though). Hence, there seems to be evidence that continuing firms only partly benefit from the growth of the industry when the industry is in an expanding stage. However, in contracting industries, continuing firms perform relatively well in terms of their aggregate sales, despite the fact that the entrants are on average not significantly less productive as incumbents.

Hypothesis IV.1 *For each period, large firms are more productive than small firms.*

The correlation coefficient between firm size (in terms of the log of sales) and productivity (real value added per employee) is equal to 0.40 for expanding industries, and 0.41 for shakeout industries, with both *p*-values far less than 0.01.[83] Hence, this finding supports the hypothesis that in every period large firms are more productive than small firms.

4.3.4 Conclusion

Let us give a short summary of the results of this section. The results of testing the first three hypotheses show that the process of entry is in line with Klepper's propositions. Annual entry rates, survival rates and cumulative entry rates are higher in expanding industries than in industries experiencing a shakeout of firms. Apparently, for entrants it becomes more difficult to survive and grow as the industry matures. The model's explanation for this is that the advantage of incumbents over entrants grows over time. However, as the results of testing hypothesis I.5 show, this is not reflected by a higher productivity of continuing firms.

No evidence was found for the hypothesis I.4: on average changes in market shares are very small for continuing firms, and about half of them actually lose market share. Hypothesis II.1 was rejected as well: in expanding industries a number of exiting firms are replaced by a higher number of entrants, but the latter are not significantly smaller than the former. Further, in the group of shakeout industries the Herfindahl-index does not increase significantly, suggesting that large firms do not grow at the expense of small firms. Hypothesis III.1 was accepted, because a positive correlation was found between the growth of market share of continuing firms (in expanding and shakeout industries) and their average profit rates. Hypothesis III.2 could not be accepted, since on average the profit rates between the group of expanding industries and shakeout industries do not significantly differ. Hypothesis III.3 was rejected, because in expanding industries continuing firms on average lose market share, whereas in industries experiencing a shakeout they gain on average. The final hypothesis (IV.1) was accepted, because a fairly strong positive correlation was found between labour productivity and the size of the firm (in terms of output), suggesting that large firms are indeed more productive than small firms.

---

[83]A higher capital intensity of large firms might also explain their higher labour productivity. However, using investment in fixed assets over sales as a proxy for capital intensity, no positive correlation was found between the log of sales and 'capital intensity'.

Returning to the three questions posed in section 4.3.1, the analysis of this section has shown that the answer to the first question is affirmative: industries in different stages show different regularities. The second question (are the observed regularities in line with the regularities predicted by the model?) cannot be answered so easily, because the evidence is mixed. On the one hand, no (or only weak) evidence was found for the hypotheses that focused on the expansion of market shares of continuing firms, or on the differences of profit rates between expanding and shakeout industries. On the other hand, the test results of hypotheses I.1 to I.3 seem to confirm Klepper's prediction of a growing advantage of incumbents over entrants, since the results suggest that is it more difficult for small firms to enter, survive, and grow in industries experiencing a shakeout. Further, we have found evidence that a shakeout of firms is not caused by an increase in exit rates, but rather by a significant reduction in the number of firms entering the industry. This finding is again consistent with the model by Klepper. Finally, large firms were found to be more productive than small firms. Hence, to some extent the observed regularities are in line with Klepper's predictions.

However, no evidence was found for the part of Klepper's model that actually explains the emerging regularities of new, technologically progressive industries. It is dynamic increasing returns to technological change that drives Klepper's model, but on average incumbent firms were not found to be more productive or profitable than entrants when mature industries are considered. Therefore, the answer to the third question seems to be no: although industries in different stages show different regularities, and although the observed (differences in) regularities are to some extent in line with Klepper's model, they cannot generally be explained by dynamic increasing returns to technological change. However, the significant differences between the groups of industries in different stages that were observed for many of the variables, indicate that distinguishing industries on the basis of their evolutionary stage still makes sense. Furthermore, there is some evidence for a growing advantage of incumbents over entrants, even though incumbents are not significantly more productive than entrants.

4.4 COMBINED ANALYSIS

Since there are a number of variables that are theoretically explained by both the framework of technological regimes and the product life cycle, it might be interesting to investigate empirically the extent to which these two approaches can simultaneously account for differences in these variables across industries, and to see whether there is any interaction between them. The most appropriate way to investigate this is to run regressions in which the independent variables are dummies denoting the technological regime and the evolutionary stage of the industries. An interesting opportunity of applying such a regression analysis is that we can now also include those industries that were left out in the previous sections (e.g., the 28 industries that were omitted in the analysis of section 4.2 because their technological regime was unknown), and see whether these industries have some special properties in common. Therefore, all 106 industries present in the SN database will be included in the regression analysis that follows. Table 4.21 shows how the technological regimes and the evolutionary stage are distributed over the industries.

In the regressions, the following six dummies are always included in the analysis: *SM-II*, *Unknown regime, Shakeout*, *Consolidating*, *Contracting* and *'Other' stage*. Interaction dummies are included on the basis of a stepwise selection method. This method maximises the number of interaction dummies included under the condition that each interaction dummy is significant at the 10 percent level. Because there are no observations for the dummies (*SM-II x Consolidating*) and for (*SM-II x Contracting*), at most six interaction dummies can be included. The group of industries that are both SM-I and in the expanding stage is thus taken as the reference group in this

Table 4.21 Distribution of Technological Regimes and Evolutionary Stages over the Industries

		Technological regime			
		Schumpeter Mark I	Schumpeter Mark II	Unknown regime	Total
Evolutionary stage	Expanding	27	12	12	51
	Shakeout	13	4	11	28
	Consolidating	5	0	1	6
	Contracting	14	0	3	17
	'Other' stage	1	2	1	4
	Total	60	18	28	106

regression analysis. Table 4.22 to 4.25 lists the results of the regression analyses for the selected industry-level variables.[84]
Regarding the regime dummies, table 4.22 shows that the SM-II dummy is significant and with the expected sign for the first three variables, and insignificant for the profit and investment margins and the survival rates. Hence, except for the profit margin, these observations corroborate the findings of section 4.2 regarding these variables: the share of small firms is significantly lower, and concentration levels and median firm sizes are significantly higher in SM-II industries.

The group of industries with unknown regimes shows a lower share of small firms and higher profit and investment margins. With regard to the evolutionary stages, the dummies denoting shakeout, consolidating and 'other' stage are not found to be statistically significant for any of these six variables, with the notable exception of survival rates in shakeout industries. In line with the results of section 4.3, survival rates are significantly lower in industries experiencing a shakeout. Contracting industries exhibit lower small firm shares, higher median firm sizes, and lower profit margins and survival rates.

The parameter estimates of the interaction terms highlight the group of industries with an unknown technological regime experiencing a shakeout. This group of industries has a lower share of small firms and a higher median firm size. Further, profit and investment margins are significantly lower in this group. Finally, the F-statistics show that the hypothesis that all parameter estimates (except for the constant) are equal to zero cannot be rejected at the 10 percent level for the Herfindahl-index. For the profit margin, this hypothesis cannot be rejected at the 5 percent level.

Next, table 4.23 shows the regression results of the annual entry and exit rates. In line with the results of section 4.2, the SM-II dummy is not significant in explaining annual firm entry and exit rates, but significant in explaining annual sales entry rates. However, annual sales exit rates are not significantly lower for SM-II industries (in contrast with the result of section 4.2).

The dummy indicating the group of unknown regime shows a significance pattern identical to the SM-II dummy: it is only significant in explaining annual sales entry rates. The shakeout dummy is only significantly negative

[84] All industry-level variables that are theoretically explained by the technological regime framework and Klepper's model are included. The variables measuring the share of small firms in the total population of an industry and the share of the largest four firms in total sales are left out because they closely correspond to respectively the share of small firms in total sales and the Herfindahl-index.

Table 4.22

	Share of small firms in total sales	Herfindahl-index	Median firm size	Profit margin	Investment margin	Survival rate
Constant	0.446***	0.107***	3.977***	0.088***	0.063***	0.780***
	(0.035)	(0.023)	(0.079)	(0.006)	(0.004)	(0.029)
SM-II	−0.320***	0.079**	0.617***	0.014	0.004	−0.009
	(0.056)	(0.038)	(0.129)	(0.009)	(0.007)	(0.048)
Unknown regime	−0.122**	0.033	0.106	0.016*	0.013*	−0.046
	(0.058)	(0.031)	(0.129)	(0.009)	(0.007)	(0.043)
Shakeout	0.069	−0.027	−0.119	0.009	−0.001	−0.152***
	(0.056)	(0.032)	(0.130)	(0.009)	(0.007)	(0.040)
Consolidating	−0.086	0.019	0.055	−0.005	−0.003	−0.013
	(0.094)	(0.058)	(0.200)	(0.014)	(0.011)	(0.074)
Contracting	−0.116**	0.041	0.232*	−0.017*	−0.010	−0.162***
	(0.057)	(0.038)	(0.131)	(0.009)	(0.007)	(0.054)
'Other' stage	0.037	0.029	0.358	0.005	−0.012	0.069
	(0.103)	(0.070)	(0.238)	(0.017)	(0.014)	(0.089)
Shakeout × SM-II						
Shakeout × Unknown regime	−0.223**		0.795***	−0.035**	−0.024**	
	(0.094)		(0.215)	(0.015)	(0.011)	
Consolidating × Unknown regime	0.404*				0.051**	
	(0.222)				(0.026)	
Contracting × Unknown regime						0.284*
						(0.115)
'Other' stage × SM-II						
'Other' stage × Unknown regime					−0.047*	
					(0.027)	
Adjusted R^2	0.282	0.008	0.285	0.065	0.106	0.162
F-statistic	6.15	1.15	6.98	2.04	2.38	3.89
(p-value)	(<0.01)	(0.34)	(<0.01)	(0.06)	(0.02)	(<0.01)

Note: For the dependent variable *Median firm size* we have taken its logarithm. Standard errors are in parentheses.
*Significant at the 10%-level.
**Significant at the 5%-level.
***Significant at the 1%-level.

Table 4.23

	Annual firm entry rate	Annual sales entry rate	Annual firm exit rate	Annual sales exit rate
Constant	0.100***	0.055***	0.078***	0.039***
	(0.007)	(0.005)	(0.008)	(0.006)
SM-II	−0.007	−0.023***	0.000	−0.009
	(0.012)	(0.008)	(0.014)	(0.009)
Unknown regime	−0.003	−0.015**	0.005	−0.006
	(0.010)	(0.007)	(0.011)	(0.008)
Shakeout	−0.033***	−0.011	−0.001	0.007
	(0.010)	(0.007)	(0.011)	(0.008)
Consolidating	−0.030	−0.009	−0.003	0.006
	(0.019)	(0.012)	(0.021)	(0.014)
Contracting	−0.018	−0.015*	0.028**	0.036***
	(0.012)	(0.008)	(0.014)	(0.009)
'Other' stage	0.003	−0.006	−0.014	−0.001
	(0.023)	(0.015)	(0.025)	(0.017)
Adjusted R^2	0.057	0.070	<0.001	0.118
F-statistic	2.05	2.33	0.94	3.35
(p-value)	(0.07)	(0.04)	(0.47)	(<0.01)

Note: None of the interaction dummies was statistically significant. Standard errors are in parentheses.
*Significant at the 10%-level.
**Significant at the 5%-level.
***Significant at the 1%-level.

for firm entry rates, confirming the results of section 4.3. Consolidating industries and industries in the 'other' stage do not show any significant differences regarding annual entry and exit rates. However, in contracting industries, entry rates (in term of sales) are significantly lower and exit rates are significantly higher.

Finally, note that the values for the adjusted R^2 for all four variables and the value of the F-statistic for annual firm exit rates are quite low.

With regard to cumulative sales entry and exit rates, the figures in table 4.24 confirm our findings of section 4.2 that these measures are significantly lower in SM-II industries. However, with regard to cumulative firm entry and exit rates the results suggest a significance pattern opposite to section 4.2. Here, we find that firm entry rates are significantly lower in SM-II industries, whereas cumulative firm exit rates are not significantly lower in this group of industries. The dummy denoting the group of unknown regimes is negatively significant for the cumulative sales entry rates

Table 4.24

	Cumulative firm entry rate	Cumulative sales entry rate	Cumulative firm exit rate	Cumulative sales exit rate
Constant	0.566***	0.425***	0.320***	0.310***
	(0.026)	(0.030)	(0.023)	(0.031)
SM-II	−0.118***	−0.211***	0.021	−0.183***
	(0.042)	(0.049)	(0.039)	(0.058)
Unknown regime	−0.012	−0.116***	0.050	−0.129***
	(0.042)	(0.041)	(0.038)	(0.046)
Shakeout	−0.211***	−0.137***	0.080**	0.007
	(0.042)	(0.041)	(0.038)	(0.047)
Consolidating	−0.162***	−0.105	−0.012	−0.004
	(0.064)	(0.076)	(0.058)	(0.078)
Contracting	−0.204***	−0.173***	0.162***	0.142***
	(0.042)	(0.050)	(0.039)	(0.057)
'Other' stage	−0.043	−0.032	0.038	−0.100
	(0.077)	(0.091)	(0.097)	(0.109)
Shakeout × SM-II				0.202*
				(0.112)
Shakeout	−0.123*		−0.152**	
× Unknown regime	(0.070)		(0.063)	
Consolidating				
× Unknown regime				
Contracting				0.392***
× Unknown regime				(0.122)
'Other' stage			−0.316**	
× SM-II			(0.139)	
'Other' stage				0.591***
× Unknown regime				(0.212)
Adjusted R^2	0.374	0.214	0.203	0.315
F-statistic	9.95	5.75	4.35	6.36
(p-value)	(<0.01)	(<0.01)	(<0.01)	(<0.01)

Note: Standard errors are in parentheses.
*Significant at the 10%-level.
**Significant at the 5%-level.
***Significant at the 1%-level.

and exit rates. The dummies denoting the group of shakeout and contracting industries show rather similar significance patterns: cumulative entry rates are lower and cumulative exit rates are higher in these two groups. The group of shakeout industries with an unknown technological regime shows significantly lower cumulative firm entry and exit rates.

Table 4.25

	Short-run turbulence among incumbents	Long-run turbulence among incumbents	Differences in the collective market shares of incumbents	Share in productivity growth of the entry and exit process
Constant	0.144***	0.267***	−0.128***	0.550***
	(0.009)	(0.019)	(0.024)	(0.124)
SM-II	−0.033**	0.020	0.047	−0.357*
	(0.015)	(0.031)	(0.038)	(0.196)
Unknown regime	−0.055***	−0.059*	0.050	−0.060
	(0.013)	(0.031)	(0.042)	(0.162)
Shakeout	−0.013	−0.049	0.225***	−0.308*
	(0.013)	(0.031)	(0.038)	(0.164)
Consolidating	−0.014	−0.059	0.132**	−0.172
	(0.023)	(0.047)	(0.058)	(0.338)
Contracting	0.016	0.051*	0.378***	0.170
	(0.015)	(0.031)	(0.043)	(0.205)
'Other' stage	0.022	0.122*	0.059	0.078
	(0.028)	(0.065)	(0.069)	(0.390)
Shakeout × SM-II				
Shakeout × Unknown regime		0.088*	−0.111*	
		(0.051)	(0.065)	
Consolidating × Unknown regime				
Contracting × Unknown regime			0.249***	
			(0.094)	
'Other' stage × SM-II				
'Other' stage × Unknown regime		−0.260**		
		(0.128)		
Adjusted R^2	0.174	0.108	0.569	0.042
F-statistic	4.68	2.59	18.3	1.14
(p-value)	(<0.01)	(0.01)	(<0.01)	(0.14)

Note: Standard errors are in parentheses.
*Significant at the 10%-level.
**Significant at the 5%-level.
***Significant at the 1%-level.

Finally, compared to the previous two groups of variables, the values of the adjusted R^2 and the F-statistic are much higher.

Table 4.25 shows the regression statistics of the final four industry-level variables. In line with the findings of section 4.2, short run market share turbulence among incumbents is significantly lower in SM-II industries, whereas long run turbulence is not. Further, the contribution of the entry and exit process in aggregate productivity growth is significantly lower in SM-II industries. The dummy denoting the group of industries with unknown regime is (negatively) significant in explaining both short and long run turbulence among incumbents. The group of shakeout and consolidating industries show a higher growth of the collective market share of incumbents and a lower share of the entry and exit process in aggregate productivity growth. In contracting industries, long run turbulence and the market share growth of incumbents are significantly higher, whereas for the industries in the 'other' stage only the long run turbulence among incumbents was found to be significantly higher.

For two out of the four variables in table 4.25, the group of unknown regime industries experiencing a shakeout stands out again. This interaction term is significantly positive in explaining long run turbulence among incumbents and significantly negative in explaining the collective market share growth of incumbents.

Finally, note that for the regression of the last dependent variable of table 4.25 (the share of the entry and exit process in aggregate productivity growth) the hypothesis that all parameter but the constant are equal to zero could not be rejected at the 10 percent level.

In conclusion, most of the results obtained in section 4.2 and 4.3 remain unchanged. Considering the group of industries that were already included in these two sections, only three minor changes can be observed. The first one addresses annual sales exit rates. In section 4.2, these rates were significantly lower in SM-II industries, whereas the SM-II dummy in the regression analysis of this section was not statistically significant. The second difference is that cumulative firm entry rates are significantly lower in SM-II industries in the regression analysis, whereas section 4.2 showed they were not significantly different. The third difference concerns the profit margins. In section 4.2, we showed that this variable was significantly higher in SM-II industries than in SM-I industries, whereas in the regression analysis of the present section the dummy denoting the group of SM-II industries was not significantly different.

When we look at the industries that were initially not included in the analysis, we observed the following results. First, the group of industries for which the technological regime was unknown showed some interesting properties. In this group, the share of small firms, annual and cumulative sales entry rates and turbulence among incumbents are lower, whereas profit and investment margins were found to be higher than in the reference group. Hence, for these variables the group with unknown regimes resembles the archetypical SM-II regime. However, the insignificant results for the other variables (most notably concentration levels and the contribution of the entry and exit process to aggregate productivity growth) do not correspond to the theoretical SM-II regime.

Second, the group of contracting industries showed some structural properties (low share of small firms, high median firm size) that may indicate circumstances conducive to large firms. In this group, selective forces seem to be rather strong: low profit margins, low entry and survival rates and high exit rates. Finally, the interaction terms highlighted the group of industries with an unknown technological regime experiencing a shakeout. This group has the following properties in common. Small firms have a minor share in the industry, the median size of the firm is high, and profit and investment margins are low. Regarding the dynamic properties, we observed for this group low cumulative firm entry and exit rates, but a high long run market share turbulence among incumbents. Incidentally, the other interaction terms were significant too, but none of them exhibited a systematically different pattern vis-á-vis the reference group of industries.

4.5 CONCLUSIONS

Starting from the hypothesis that differences between industries can be explained by underlying structural differences (i.e., different technological regimes) as well as by temporal differences (i.e., differences in the evolutionary stages of the industries), this chapter has shown that the data indeed provide evidence for both types of explanations. The analysis of section 4.2 strongly suggested that differences in the structural and dynamic properties of industries are closely related to the set of opportunity, appropriability, cumulativeness and knowledge conditions underlying the innovative activities in an industry. And although the model by Klepper (1996) is primarily intended to depict the evolution of technologically progressive industries, section 4.3 has shown that applying the model to (basically mature) manufacturing industries in general, i.e., without any reference to their technological progressiveness, provides some interesting results. Industries in different evolutionary stages indeed show different regularities, which are partly in line with Klepper's model. Especially differences in the entry and survival rates of industries corroborate Klepper's hypothesis of a growing advantage of incumbents over entrants. Further, we have found circumstantial evidence that a shakeout of firms is not caused by an increase in exit rates, but rather by a significant reduction in the number of firms entering the industry. This finding is again consistent with the model by Klepper. Apart from some minor differences, the regression analysis of section 4.4, which investigated the interaction between the two theoretical approaches, generally confirmed the results of section 4.2 and 4.3.

Further analyses, directly using proxies for the actual opportunity, appropriability, cumulativeness and conditions of knowledge accumulation to determine the technological regime, or the actual age of an industry together with, e.g., the growth rate of patents to determine an industry's evolutionary stage would certainly provide a stronger basis for testing whether technological regimes and industry life cycles explain the observed differences between industries. However, we believe that the analysis presented here has provided important evidence on the links between technological regimes, patterns of innovation, industry life cycles and industrial dynamics.

Appendix: List of Industries with Their Technological Regime and Evolutionary Stage

Industry	Technological regime	Evolutionary stage
Finishing of textiles	Schumpeter Mark I	Expanding
Manufacture of carpets and rugs	Schumpeter Mark I	Expanding
Manufacture of made-up textile articles, except apparel	Schumpeter Mark I	Expanding
Manufacture of other non-metallic mineral products nec*	Schumpeter Mark I	Expanding
Manufacture of glass and glass products	Schumpeter Mark I	Expanding
Iron, steel and non-ferrous metal foundries	Schumpeter Mark I	Expanding
Manufacture of tanks, reservoirs and pipe-lines	Schumpeter Mark I	Expanding
Manufacture of steel and non-ferrous metal doors, windows, walls and the like	Schumpeter Mark I	Expanding
Metal construction nec	Schumpeter Mark I	Expanding
Manufacture of metal furniture	Schumpeter Mark I	Expanding
Forging, treatment and coating of metals	Schumpeter Mark I	Expanding
Manufacture of agricultural machinery	Schumpeter Mark I	Expanding
Manufacture of machine-tools	Schumpeter Mark I	Expanding
Manufacture of machinery for packing and wrapping	Schumpeter Mark I	Expanding
Manufacture of machinery for food, beverage and tobacco processing	Schumpeter Mark I	Expanding
Manufacture of machinery for manufacturers of rubber and plastic products	Schumpeter Mark I	Expanding
Manufacture of lifting and handling equipment	Schumpeter Mark I	Expanding
Manufacture of machinery for textile and apparel	Schumpeter Mark I	Expanding
Manufacture of machinery for chemical cleaning, washing, leather and leather products, paper and paper products and printing	Schumpeter Mark I	Expanding
Manufacture of pumps, compressors, taps and valves	Schumpeter Mark I	Expanding
Manufacture of fans, refrigerating and freezing equipment	Schumpeter Mark I	Expanding
Appendage	Schumpeter Mark I	Expanding
Manufacture of machine parts nec	Schumpeter Mark I	Expanding
Manufacture of machinery nec	Schumpeter Mark I	Expanding
Manufacture of trailers and semi-trailers	Schumpeter Mark I	Expanding
Manufacture of bodies for motor vehicles	Schumpeter Mark I	Expanding
Manufacture of transport equipment nec	Schumpeter Mark I	Expanding
Manufacture of other textiles nec	Schumpeter Mark I	Shakeout
Tanning and dressing of leather	Schumpeter Mark I	Shakeout
Manufacture of luggage, handbags and the like, saddlery and harness	Schumpeter Mark I	Shakeout
Manufacture of wooden containers	Schumpeter Mark I	Shakeout
Manufacture of other products of wood, manufacture of articles of cork, straw and plaiting materials	Schumpeter Mark I	Shakeout

(Continued)

Industry	Technological regime	Evolutionary stage
Manufacture of furniture, except metal furniture	Schumpeter Mark I	Shakeout
Manufacture of cement and lime	Schumpeter Mark I	Shakeout
Manufacture of articles of concrete and cement	Schumpeter Mark I	Shakeout
Forging, pressing, stamping and roll-forming of metal	Schumpeter Mark I	Shakeout
Manufacture of metal fasteners, cables, springs and the like	Schumpeter Mark I	Shakeout
Manufacture of heating and boilers, except electrical	Schumpeter Mark I	Shakeout
Manufacture of other fabricated metal products nec	Schumpeter Mark I	Shakeout
Machine repair nec	Schumpeter Mark I	Shakeout
Carpentry and manufacture of densified wood and parquet flooring blocks	Schumpeter Mark I	Consolidating
Manufacture of ceramics	Schumpeter Mark I	Consolidating
Manufacture of basic metals	Schumpeter Mark I	Consolidating
Manufacture of tools and machinery for metallurgy	Schumpeter Mark I	Consolidating
Manufacture of bearings, gears, gearing and driving elements	Schumpeter Mark I	Consolidating
Preparation and spinning of wool fibres, weaving of wool	Schumpeter Mark I	Contracting
Preparation and spinning of cotton fibres, weaving of cotton	Schumpeter Mark I	Contracting
Manufacture of tricot and stockings	Schumpeter Mark I	Contracting
Manufacture of wearing apparel, dressing and dyeing of fur	Schumpeter Mark I	Contracting
Manufacture of footwear	Schumpeter Mark I	Contracting
Sawmilling and planing of wood, manufacture of veneer sheets, plywood, laminboard, particle board and other panels and boards	Schumpeter Mark I	Contracting
Manufacture of bricks and tiles	Schumpeter Mark I	Contracting
Manufacture of metal packings	Schumpeter Mark I	Contracting
Manufacture of machinery for petrochemical, chemical and pharmaceutical industries	Schumpeter Mark I	Contracting
Manufacture of machinery for wood and furniture	Schumpeter Mark I	Contracting
Manufacture of engines and turbines, except aircraft, vehicle and cycle engines	Schumpeter Mark I	Contracting
Manufacture of office machinery	Schumpeter Mark I	Contracting
Manufacture of weighing machinery and domestic appliances, except electrical	Schumpeter Mark I	Contracting
Building and repairing of ships and boats	Schumpeter Mark I	Contracting
Manufacture of machinery for mining, construction, building materials and metallurgy	Schumpeter Mark I	'Other' stage
Manufacture of synthetic resin	Schumpeter Mark II	Expanding
Manufacture of dye-stuffs and colouring matters	Schumpeter Mark II	Expanding

(*Continued*)

Industry	Technological regime	Evolutionary stage
Manufacture of chemical raw materials nec	Schumpeter Mark II	Expanding
Manufacture of paints, varnishes and similar coatings, printing ink	Schumpeter Mark II	Expanding
Manufacture of pharmaceuticals and medicinal chemicals	Schumpeter Mark II	Expanding
Manufacture of other chemical products nec	Schumpeter Mark II	Expanding
Manufacture of plastic products	Schumpeter Mark II	Expanding
Manufacture of electric motors, generators and transformers	Schumpeter Mark II	Expanding
Manufacture of electricity distribution and control apparatus	Schumpeter Mark II	Expanding
Manufacture of other electrical equipment nec	Schumpeter Mark II	Expanding
Manufacture of motor vehicles	Schumpeter Mark II	Expanding
Manufacture of parts and accessories for motor vehicles	Schumpeter Mark II	Expanding
Manufacture of soap and detergents, cleaning and polishing preparations, perfumes and toilet preparations	Schumpeter Mark II	Shakeout
Manufacture of chemical pesticides	Schumpeter Mark II	Shakeout
Manufacture of rubber products	Schumpeter Mark II	Shakeout
Manufacture of motorcycles and bicycles	Schumpeter Mark II	Shakeout
Manufacture of fertilisers	Schumpeter Mark II	'Other' stage
Manufacture of insulated wire and cable	Schumpeter Mark II	'Other' stage
Processing and preserving of fish and fish products	Unknown regime	Expanding
Manufacture of bakery products	Unknown regime	Expanding
Manufacture of other food products nec	Unknown regime	Expanding
Manufacture of malt liquors and malt	Unknown regime	Expanding
Manufacture of other articles of paper and paperboard	Unknown regime	Expanding
Manufacture of corrugated paper and paperboard	Unknown regime	Expanding
Offset printing	Unknown regime	Expanding
Chemigrafical and fotolithografical firms	Unknown regime	Expanding
Other printing	Unknown regime	Expanding
Publishing of books	Unknown regime	Expanding
Other publishing	Unknown regime	Expanding
Manufacture and repair of aircraft	Unknown regime	Expanding
Production, processing and preserving of meat and meat products	Unknown regime	Shakeout
Manufacture of grain mill products	Unknown regime	Shakeout
Manufacture of vegetable and animal oils and fats	Unknown regime	Shakeout
Processing and preserving of fruit and vegetables	Unknown regime	Shakeout
Manufacture of sugar, cacao, chocolate and sugar confectionery	Unknown regime	Shakeout
Manufacture of prepared animal feeds	Unknown regime	Shakeout
Manufacture of soft drinks	Unknown regime	Shakeout

(*Continued*)

Industry	Technological regime	Evolutionary stage
Manufacture of tobacco products	Unknown regime	Shakeout
Manufacture of paper and paperboard	Unknown regime	Shakeout
Printing of newspapers	Unknown regime	Shakeout
Publishing of newspapers	Unknown regime	Shakeout
Bookbinding	Unknown regime	Consolidating
Manufacture of dairy products	Unknown regime	Contracting
Distilling, rectifying and blending of spirits; ethyl alcohol production	Unknown regime	Contracting
Printing of books	Unknown regime	Contracting
Publishing of periodicals	Unknown regime	'Other' stage

*not elsewhere classified.

5. TECHNOLOGICAL DIFFUSION PATTERNS AND THEIR EFFECTS ON INDUSTRIAL DYNAMICS

5.1 INTRODUCTION

Both the technological regime framework and the industry life cycle approach are able to explain variations in industry structures and dynamics. The previous chapter has shown that opportunity, appropriability, cumulativeness and knowledge conditions underlying the industrial innovative process, as well as the evolutionary stage of the industrial life cycle affect the structural and dynamic properties of an industry. However, as we have mentioned in section 3.5, in our view these theories still ignore a number of crucial elements. First of all, models of product life cycles generally focus on the emergence and evolution of only one product and its associated technology. However, in many industries we observe that firms repeatedly introduce or adopt new product technologies that replace the older ones. Second, both the technological regime and the industry life cycle approach do not explicitly consider differences in the technological properties of the goods produced by the industries. Finally, in models on technological regimes and on industry life cycles the growth of a firm is generally determined by its relative (technological) performance. However, empirical studies on firm growth do not provide much evidence supporting such a relationship. Most of these studies suggest that the size of a firm generally follows a random walk with a declining positive drift.[85]

In this chapter we will introduce a model on industry dynamics that attempts to include these three elements. As in Shy (1996), the degree of

[85]See Geroski (1998).

substitutability between the quality and the network size of a technology and the degree of compatibility of succeeding technologies are the key determinants of the simulation model presented here. However, Shy (1996) mainly limits his focus to the demand side, as he investigates how varying consumer preferences over technology advance and network size effects the timing and frequency of new technology adoption. Our focus will be on the relation between the demand side and the supply side. Given variations in purchaser preferences over quality and network sizes, and different degrees of compatibility between succeeding technologies, we will investigate how the resulting differences in the timing and frequency of new technology adoptions by purchasers effect the dynamics of the population of supplying firms. Furthermore, we will investigate whether these effects are different under various technological regimes.

The structure of the chapter is as follows. The next section presents the conceptual basis of the model. Based on Shy (1996), we will explain how variations in consumer preferences over quality and network sizes of product technologies and different degrees of compatibility between succeeding technologies affect the timing and frequency of new technology adoptions. Further, we will elaborate on some theoretical and empirical issues concerning firm growth. Section three formally presents the model, of which the simulation results will be analysed in section four. Section five focuses on how these results are effected when different technological regime conditions are considered. Section six concludes this chapter.

5.2 THE DIFFUSION OF NEW PRODUCT TECHNOLOGIES

As mentioned in the introduction, one of the objectives of the industry model presented in this chapter is to include a repeating process of new technology adoptions. An interesting option of including such a process is that, by varying the parameters determining the diffusion patterns, the model can be used to investigate whether differences in the diffusion of new technologies affect the structural and dynamic properties of an industry. This would be an important result, because if the industrial properties are indeed related to the diffusion patterns, our model may provide an additional explanation for the observed structural and dynamic differences between industries.

Obviously, we will first need a theoretical framework that explains how differences in the diffusion patterns of new product technologies may arise.

Although there is a considerable amount of literature on modelling diffusion dynamics[86], in our view the model by Shy (1996) provides the most appropriate framework for analysing the determinants of diffusion patterns. In this model, the degree of substitutability between the quality and the network size of a technology and the degree of compatibility of succeeding technologies are the key variables determining the timing and frequency of new technology adoptions. In what follows we will briefly explain the logic behind Shy's model.

5.2.1 Modelling Diffusion Dynamics: Shy's Approach

Shy (1996) describes the consumer dynamics within an overlapping generations framework. In the model, the generation of entering consumers chooses whether to purchase a certain product based on an old technology already used by an older generation of consumers or whether to purchase the product based on the new technology with a higher quality. The young generation chooses the new technology if the utility from the high quality technology, combined with the size of the network associated with the new technology, overtakes the utility from the old technology with its associated network size. The size of the network of the new technology is the sum the population size of the young generation and a certain percentage of the old generation of users. This percentage is determined by the degree of compatibility between the old and new technology. Hence, the higher the compatibility, the larger the network size associated with the new technology will be. Shy (1996) then shows that a decrease in the degree of compatibility between new and old technologies will increase the duration of each technology. Further, by varying the degree of substitution between the quality and the network size of a technology, he shows that the duration of each adopted technology is lower and the frequency of technology adoptions is higher the more consumers value quality and network size as substitutes rather than complements.

Hence, his focus on consumer preferences helps us understand "*...why technology is replaced more often in some industries than in others...*" (Shy, 1996, p. 786). He also asserts that his model is general enough to capture a variety of market structures. That is, he shows that both a persistent monopoly, as well as a more competitive market structure with the entry of

[86]See Stoneman (1991) for an overview of these models.

a new firm whenever a new technology becomes available[87], is consistent with his model. The aim of our model is to further elaborate on the evolution of the supply side of industries experiencing repeated adoption of new technologies. Although previous models on technological change and industry evolution have investigated issues such as the evolution of the firm population[88] or the diffusion process of (subsequent) innovations[89], they have not explicitly linked these two processes such that causality runs from the latter to the former. Hence, these models have not analysed the effect of technology diffusion patterns on the dynamics of the firm population. The model presented in this chapter attempts to fill this hiatus.

In conclusion, the model by Shy (1996) provides the theoretical basis of the first two distinguishing elements that are included in our model on industry dynamics (i.e., the repeated adoption of new product technologies, and differences in technological properties of the goods produced). The next subsection will focus on the process governing firm growth that constitutes the third distinguishing element of our model.

5.2.2 A Note on Firm Growth

In virtually all evolutionary models on industrial dynamics some type of replicator dynamics is employed to depict the evolution of the size of a firm. The basic assumption underlying this mechanism is that the selection process allocates market share from the less efficient firms to the more efficient ones. Hence, in its simplest form this replication concept states that the growth of a firm over a certain period is proportional to its relative competitiveness. However, there are a number of reasons why in practice such a relationship may not be very strong. Let us briefly summarise these reasons.

The growth or decline of a firm is ultimately a managerial decision: the management decides on how much to invest and how many workers to hire or lay off. Many factors may influence the final outcome of this decision. Past performance is certainly one of them. Besides creating the necessary funds, high profits in the past are a signal of a firm's competitiveness,

---

[87]Shy (1996) assumes that the new firm is endowed with a one period patent right on the new technology, allowing it to (temporarily) charge a monopoly price.

[88]See, e.g., Nelson and Winter (1982), Winter (1984) and Dosi et al. (1995).

[89]See, e.g., Davies (1979), Iwai (1984a, 1984b), Reinganum (1981), Silverberg and Lehnert (1993), Silverberg and Verspagen (1994a, 1994b).

creating confidence among the management and the potential investors. However, high profits may also indicate a lag in mobilising effective competition, reflecting a windfall gain from being properly positioned to take advantage of a change in level or character of demand. In this last case the high profits may have resulted from mere chance, and are not likely to be as persistent as profits resulting from having been superior to competitors.

But probably more important than a firm's past competitiveness in the growth decision are the expectations of a firm regarding the state of the economy, the condition of the industry, a firm's own performance, et cetera. There are many reasons why firms may have different expectations. First of all, the prospects may differ between industries. When an industry is expected to grow rapidly, a firm will be more inclined to expand than when prospects are less optimistic for the industry. Second, firms within the same industry could have different information sets on which the decision to grow is based. Third, even if firms would have the same information, still they may perceive and interpret it differently, leading to different expectations.

Naturally, different expectations lead to different decisions. Moreover, firms may have different ambitions regarding their (ultimate) sizes or market shares. Some firms may indeed be driven by enormous ambitions and try to capture the total market as much as possible. But other firms may be less ambitious. Their aim could be to acquire a certain amount of profits, and if this goal is reached at a certain size they may decide to keep the size approximately fixed. Perhaps the desire to grow is present latently, but if it is not strong enough no serious attempts will be made to fulfil this desire. Finally, the situation on the markets for labour and capital goods may differ across industries. A shortage of labourers with skills necessary for a specific firm probably hinders a firm's desire to expand.

Combining the potential differences in expectations, ambitions and input markets makes anticipating the growth paths of firms very difficult. As Geroski (1998) argues, the growth of a firm may very well be understood, but also be hard to describe or predict with any precision. Perhaps for these reasons econometric work on the growth of firms usually find that firm size essentially follows a random walk. Only some significantly negative effects of initial size and age on corporate growth are frequently observed, but usually these effects are not very large (Geroski, 1998).

As shown in chapter 2, in Dutch manufacturing these regularities are also observed. We have also found that initial size and firm age (at least the proxy we have used for the age of entering firms) have a significant negative effect on firm growth. But the main question here concerns the extent to which the relative competitiveness of firms can explain their growth rates. In order to answer this question we will run two regressions for all continuing firms, in which we will employ the relative labour productivity (*RelProd*) of a firm to denote its relative competitiveness. For a given year this variable is calculated as a firm's individual labour productivity (value added per employee) divided by the weighted industry average for that year. The first regression we will run takes the actual annual growth rate of the firms as the dependent variable, the second one takes the average growth rate of the firm over the period 1978–1992. Hence, we have:

$$\ln(empl_{t+1}) - \ln(empl_t) = \beta_1 + \beta_2 \ln(RelProd_t) + \varepsilon, \tag{5.1}$$

$$[\ln(empl_{1992}) - \ln(empl_{1978})]/14 = \beta_1 + \beta_2 \ln(RelProd_{1978}) + \varepsilon, \tag{5.2}$$

The following table lists the results of the regression analyses.

Although the parameter estimates for relative labour productivity are significant in both regressions, they do not explain more than two percent of the total variance in firms' growth rates. This clearly confirms that the patterns of firm growth are only weakly influenced by selective pressures. We conclude therefore that relative competitiveness only plays a limited role in explaining the growth patterns of firms. In our view, this result is enough to abstain from an attempt to model the growth process of a firm on the basis of past or current performance levels.

In the model that follows, we will thus refrain from employing some type of replicator dynamics in modelling firm growth. Instead, we will model the evolution of the size of a firm as following a random walk, however with a declining positive drift. In this way we assure that the process governing firm growth is *a priori* consistent with the stylised fact that growth rates are negatively correlated to the firm's age. Of course, the relative competitiveness of a firm will matter in the model, but only in determining the probability of survival. Therefore, in this model it is really a matter of *survival* of

Table 5.1 Regression Analyses of Firm Growth on Relative Productivity

Regression	β_1	β_2	Adjusted R^2
(5.1)	0.005 (0.001)	0.048 (0.002)	0.017
(5.2)	0.004 (0.001)	0.012 (0.002)	0.012

Note: Standard errors in parentheses. All parameter estimates are significant at the 1%-level.

the fittest, as opposed to *expansion* of the fittest. Modelling firm growth in this way also allows us to investigate whether replicator dynamics are essential in evolutionary models on industry dynamics, or whether random firm growth can lead to realistic results as well.

After having described the two most distinguishing features of the model, i.e., its focus on the repeated adoption of new technologies in relation to the dynamics of the firm population, and the modelling of firm growth as an essentially random process, we now turn to the technical details of the model.

5.3 THE MODEL

Consider an industry where in each discrete time period t, $t = 0, 1, 2, \ldots$, the firm population consists of $N(t)$ firms. All firms in the industry are producing a certain product that is defined by its functional characteristics. An essential assumption in the model is that the function the product performs can be based on different technologies. For instance, both the standard compact cassettes as well as the compact disc (CD) are sound recording media, however analogue recording technology underlies the compact cassette, whereas a CD is recorded using digital technology. However, our notion of a product also extends to producer or capital goods. An example here could be industrial lathes, which can be manually operated or operated by computer numeric control (CNC) technology.

Every period a random number of new firms enter the industry according to a Poisson process[90] with arrival rate ρ_{ent}. At birth, each firm i is endowed with a firm-specific organisational competitiveness level λ_i, a product technology Ψ, and a size s_i. The organisational competitiveness level λ_i is a random genotype variable[91] that sets for each firm a potential limit to its actual competitiveness, creating some (initial) heterogeneity among firms with regard to their organisational capabilities. As mentioned, this variable may limit the firm's actual competitiveness, but whether it actually does depends on its technological competitiveness that is calculated as follows.

[90]For practical reasons, we have adopted the Poisson process here and approximated it by five hundred Bernoulli trials every period. For analytical convenience, the arrival rate is kept constant over the simulation period.

[91]This variable is generated as follows. Let $x \sim \mathrm{N}(\mu_\lambda, \sigma_\lambda)$. The $\lambda_i = \max\{0; x\}$ if $x \leqslant \mu_\lambda$, and $\lambda_i = \max\{0; 2\mu_\lambda - x\}$ if $x > \mu_\lambda$.

5.3.1 Competitiveness of Firms

Assume that at every period K product technologies are available. At birth, every firm is randomly endowed with a technology Ψ ($\Psi = 1, 2, \ldots, K$), such that the probability of receiving a given technology is equal to $1/(K)$. These technologies are ranked according to their intrinsic quality level Q_Ψ, such that $Q_K > Q_{K-1} > \cdots > Q_1$. Further, there is a class of old technologies $\Psi = 0$ that all have an intrinsic quality level Q_0. Every β_1 periods a pioneering entrant or incumbent introduces a new, intrinsically better product technology that has become available due to exogenous technological change. This introduction causes all technologies to drop one level in their intrinsic quality. Hence, the newly introduced technology becomes K (the technology with the highest quality level Q_K), and $\Psi = 1$ becomes part of the class of old technologies $\Psi = 0$ and degrades to the intrinsic quality level Q_0. Although our model embodies intra-firm technology diffusion (firms can employ more than one product technology simultaneously) we will first explain the evolution of some essential variables for a single-technology firm.

A firm's technological competitiveness $TC_{i,\Psi}(t)$ depends on the intrinsic quality Q_Ψ of the product technology it is applying and the total share Γ_Ψ of this technology in the industry in the following way:

$$TC_{i,\Psi}(t) = \alpha\Gamma_\Psi(t) + (1 - \alpha)Q_\Psi(t), \tag{5.3}$$

where $0 \leqslant \alpha \leqslant 1$, and $Q_0 \leqslant Q_\Psi \leqslant 1$. The parameter α is essential here, as it determines the strength of the network externalities on the demand side. The higher α, the more the total market share of a technology determines the firm's technological competitiveness.

Combining the organisational competitiveness λ_i with the technological competitiveness $TC_{i,\Psi}(t)$ gives the potential competitiveness $PC_{i,\Psi}(t)$ of a firm, which is:

$$PC_{i,\Psi}(t) = \min\{TC_{i,\Psi}(t); \lambda_i\}. \tag{5.4}$$

Hence, a firm's potential competitiveness is either bounded by its organisational or its technological competitiveness. We could have modelled organisational and technological competitiveness as (imperfect) substitutes, but this would have implied that, e.g., a firm with a very low level of organisational competitiveness may still survive as long as it has a high level of technological competitiveness. We believe that such a situation is not realistic, as firms will always need a certain level of organisational skills in order to manage the manufacturing and selling of their products. Furthermore, by modelling organisational and technological competitiveness as complementary, we exclude in advance the awkward possibility that the organisational and technological skills of the firms will be negatively correlated in the simulation results.

Finally, a firm's actual competitiveness $C_{i,\Psi}(t)$ evolves according to the following moving average process:

$$C_{i,\Psi}(t) = \theta C_{i,\Psi}(t-1) + (1-\theta)PC_{i,\Psi}(t), \tag{5.5}$$

where $0 \leqslant \theta \leqslant 1$, and $C_i(t) = \beta_2 \lambda_i$ for all firms that enter at period t. The parameter β_2 puts an entrant at an initially disadvantageous and possibly even hazardous position. To some extent such an entry process corresponds to Jovanovic (1982). Even firms with very low competitiveness levels may decide to enter the industry, simply because they do not know their true competitiveness prior to their entry. Only by actually entering they can gather some evidence regarding their real capabilities, which may subsequently eventuate in a rapid exodus of entrants with low competitiveness levels. It is also consistent with the empirical evidence on the entry process.[92]

5.3.2 Exit Rules

If the actual competitiveness is below a certain fraction Φ_L of the industry average $\bar{C}$, or if size drops below the minimum level $\hat{s}$, a firm dies with probability one, a higher productivity level reduces the probability of exiting P_{exit} (t). Survival is guaranteed for the next period if relative competitiveness exceeds an upper level Φ_H. Hence, we have

$$P_{exit,i}(t) = 0 \quad \text{if } C_{i,\Psi}(t) \geqslant \Phi_H \bar{C}(t) \tag{5.6a}$$

$$P_{exit,i}(t) = \left(\frac{\Phi_H \bar{c}(t) - c_i(t)}{\Phi_H \bar{c}(t) - \Phi_L \bar{c}(t)}\right)^{\beta_3} \quad \text{if } \Phi_L \bar{C}(t) < C_{i,\Psi}(t) < \Phi_H \bar{C}(t) \tag{5.6b}$$

$$P_{exit,i}(t) = 1 \quad \text{if } C_{i,\Psi}(t) \leqslant \Phi_L \bar{C}(t), \text{ or if } s_i(t) \leqslant \hat{s}. \tag{5.6c}$$

[92] See Caves (1998) for an extensive overview.

These exit rules can be interpreted as a mixture of voluntary and forced exit. If relative competitiveness is lower than Φ_L, or if firm size drops below the minimum level $\hat{s}$, firms go bankrupt and are thus forced to exit. However, for those firms that observe that their relative competitiveness lies between Φ_L and Φ_H, the exit decision is voluntary. Depending on their aspiration level, some of them may decide to continue, whereas others may voluntarily leave the industry and perhaps try their chances elsewhere.

5.3.3 Evolution of Firm Size

As mentioned before, all firms are initially endowed with a fixed size s_i. We interpret this size as a firm's sales capacity and assume for convenience that firms always operate at full capacity. All surviving firms grow each period according to a random process, however their mean growth rates asymptotically reach zero as they mature. The process governing firm growth is

$$\frac{s_i(t+1)}{s_i(t)} = 1 + [\beta_4 e^{-\beta_5 a_i(t)} + \chi(e^{-\beta_5 a_i(t)} + \beta_6)], \tag{5.7}$$

where $a_{i,t}$ denotes the age of firm i at t. The variable χ is randomly drawn from a normal distribution with mean $\mu_\chi = 0$ and variance σ_χ, β_4 sets the average growth rate of firms at the age of zero. This growth rate gradually declines as the firm matures, a process of which the pace is determined by β_5. Finally, β_6 assures that even at a high age the size of a firm is still subject to random shocks.

5.3.4 Imitation

The description of the model, so far, has only considered firms employing one product technology. But, as mentioned, the model also allows for firms employing several technologies simultaneously. Let $\Psi = A$ denote the firm's *intrinsically best* technology. As long as $A < K$, a firm may have the opportunity to imitate an intrinsically better technology. Every period, firms randomly receive an imitation draw according to a Poisson process[93] with arrival rate ρ_{im}. Receiving an imitation draw means that the firm acquires the knowledge of employing one intrinsically better technology. This process

[93] Again approximated by Bernoulli trials.

is arranged such that on average a firm with a given market share[94] z_i would receive ρ_{im} $[z_i+\beta_7(1-z_i)]$ imitation draws in every β_1 periods, where $0 \leqslant \beta_7 \leqslant 1$. The parameter β_7 sets the inequality between firms with different sizes with regard to receiving an imitation draw. If $\beta_7=1$ all firms have equal probabilities to imitate, if $\beta_7=0$ the probability to imitate is proportional to a firm's market share.

If an imitation draw is received, the probability of acquiring the knowledge of given other product technology is equal to $1/(K-A)$.[95] Every firm that has obtained an opportunity reallocates every period a share $\omega_i(t)$ of its total capacity from its worst available technology $\Psi=L$ (i.e., the technology with the lowest technological competitiveness) to its best available technology $\Psi=H$ (i.e., the technology with the highest technological competitiveness).[96] The size of this reallocation share $\omega_i(t)$ is determined by:

$$\omega_i(t) = \beta_8[TC_{i,H}(t) - TC_{i,L}(t)](1+\eta), \tag{5.8}$$

where β_8 is a system parameter ($0 \leqslant \beta_8 \leqslant 1$), and η is a random variable[97] drawn from a normal distribution with mean $\mu_\eta = 0$ and variance σ_η. Hence, on average the share that is reallocated increases with the difference in the technological competitiveness between the worst and the best available product technology.

Whenever capacity is reallocated, part of it gets lost because of adjustment costs. If we denote $s_{i,L}$ as the capacity allocated to the worst product technology, $s_{i,H}$ as the capacity allocated to the best product technology, s_i as the total capacity of the firm (i.e., $\sum_{\psi=L}^{H} s_{i,\psi}$), and Δs_i $(=s_i(t+1)-s_i(t))$ as the capacity growth of the firm, we have:

$$s_{i,H}(t+1) = s_{i,H}(t) + \beta_9 \min\{\omega_i(t)s_i(t); s_{i,L}(t)\} + \Delta s_i \quad \text{if } \Delta s_i > 0, \tag{5.9a}$$

[94]By market share we mean the firm's share in the total capacity of the industry. Hence, we have $z_i(t) = s_i(t)/\sum_{N(t)} s_i(t)$.

[95]The randomness of this process essentially reflects a bound to the agents' rationality, combined with some degree of technological uncertainty. Hence, the combination of these elements may lead to erroneous decisions of firms with regard to the allocation of their imitation efforts.

[96]Note that what we call here the 'best' technology is not necessarily the technology with the highest intrinsic quality.

[97]Again, bounded rationality and technological uncertainty justify the randomness of this process.

$$s_{i,H}(t+1) = s_{i,H}(t) + \beta_9 \max\{\omega_i(t)s_i(t); s_{i,L}(t) + \Delta s_i; 0\} + \min\left\{\sum_{\psi=L}^{H-1} s_{i,\psi} + \Delta s_i; 0\right\} \quad \text{if } \Delta s_i \leqslant 0, \tag{5.9b}$$

$$s_{i,L}(t+1) = \max\{\min\{s_{i,L}(t) + \Delta s_i - \omega_i(t)s_i(t); s_{i,L}(t) - \omega_i(t)s_i(t)\}; 0\}, \tag{5.10}$$

where $0 \leqslant \beta_9 \leqslant 1$. This parameter arranges the fraction of the transferred capacity that gets lost whenever reallocated. Expression (5.9a) says that if the capacity of the firm grows, it allocates first of all its growth to the best technology. Second, the firm reallocates capacity from the worst technology according to the amount determined by (5.8). If this amount is not available, it takes away all the capacity that remained for the worst technology[98], and adds it to the capacity of the best technology.

Expression (5.9b) deals with cases of negative growth of total capacity. In such a case the firm first withdraws the change in capital from the worst technology. If that is not sufficient, it will subsequently take away capacity from the second worst, the third worst, et cetera. Only if even the capacity of the second best technology $H-1$ is divested, the firm will necessarily have to withdraw the remaining part of its total capacity decline Δs_i from its best technology. If, however, $s_{i,L}(t)$ is still positive after subtracting (formally adding) Δs_i, the firm will reallocate again capacity from the worst to the best technology, possibly bounded by $s_{i,L}(t) + \Delta s_i$. Expression (5.10) gives the amount of capacity available for the worst technology after having gone through the process of reallocation.

With regard to the competitiveness and survival probabilities, we in fact regard a firm employing more than one product technology as a 'mother' firm consisting of several subfirms, each of them employing one technology. The actual competitiveness of each subfirm still evolves according to (5.5). Hence, the organisational competitiveness of the mother firm still applies to all the subfirms. Given the reallocation rules, it may happen therefore that a firm shifts part of its capacity to a certain technology because of its higher technological competitiveness, whereas the actual competitiveness derived from this technology is still bounded by the organisational competitiveness. This can be justified in two ways. First, we could argue that in this way a

[98]For simplicity, we assume that in such a case the firm only in that period does not consider reallocation from the second worst to the best technology.

firm protects itself for the long run. Somewhere in the future the technological competitiveness of the worst technology may fall below the organisational competitiveness if new technologies are introduced, in which case the technological competitiveness will be binding. In order to avoid this a firm may decide already now to transfer some capacity to the best technology. Second, we could assume that a firm only has fuzzy information with regard to its competitiveness. For instance, it may erroneously think it could perform better by switching to a better product technology.

When a firm is employing several technologies simultaneously, the actual competitiveness of the whole firm i is the weighted average of the actual competitiveness levels of all subfirms:

$$C_i(t) = \sum_{\psi} \left[\left(\frac{s_{i,\psi}(t)}{s_i(t)} \right) C_{i,\psi}(t) \right] \tag{5.11}$$

For the mother firm and all the subfirms the exit conditions expressed in (5.6) apply. In case of exit a subfirm, its capacity is totally lost. Naturally, in case the mother firm dies, all subfirms cease to exist as well.

It was already mentioned that every β_1 periods a pioneering entrant or incumbent introduces a completely new technology. When this happens, all technologies that a given firm is employing drop one level in their intrinsic quality. Further, the subfirm employing $\Psi = 1$ is merged with the subfirm employing the class of old technologies $\Psi = 0$. Hence, in that period $s_{i,1}(t)$ is added to the capacity allocated to $\Psi = 0$. With regard to the actual competitiveness of the subfirm employing $\Psi = 0$ we calculate a size weighted average of the actual competitiveness levels of the merging subfirms for that period. Since it is unlikely that, whenever a firm imitates a product technology, it could immediately fully benefit from the imitated technology, we assume that $C_{i,\Psi}(t) = \beta_{10}\ \lambda_i$ (where $0 \leqslant \beta_{10} \leqslant 1$) for all firms that imitate technology Ψ at t.

5.4 SIMULATION RESULTS

Similar to the notion of technological regimes, we introduce the notion of technology adoption regimes to classify cases with different levels of compatibility between old and new product technologies, and different degrees of substitution between the quality and the network size of a technology.

In contrast with Shy (1996) however, our concept of compatibility between technologies is not related to the notion of overlapping generations of users. In the interpretation of our model, a purchaser, repeatedly buying a given product, is more willing to switch to a newer product technology if its compatibility with the old technology is higher. To use again the example of sound recording media, a consumer that wants to replace his old analogue cassette-player would be more willing to buy a digital compact cassette (DCC) player than a CD-player, simply because his previously recorded tapes can also be played on the DCC-player. Of course, in real life there are many other considerations involved, but purely for the sake of compatibility this consumer would find a DCC player easier than a CD player. Or let us consider a case in which a firm is considering to switch from a manually operated lathe to a CNC operated one. If the CNC lathe requires worker skills that are very different from the skills necessary for manually operated lathes, the much more advanced CNC lathe is quite incompatible with the old fashioned manually controlled workbench, which could seriously hinder the adoption of this new technology.

In the model, we will simulate different compatibility levels by varying the differences in the intrinsic quality levels Q_{Ψ} between succeeding product technologies. High compatibility between a new and an old technology then implies a high difference in their intrinsic quality levels. For simulating various degrees of substitution between the quality and the network size of a technology, we will use the parameter α, which sets the relative importance of a technology's share in the market.

We will simulate three different adoption regimes, in which the number of product technologies available is equal to three ($K = 3$). The first one will be labelled 'quality regime': in this regime, quality and network size are perfect substitutes. This is arranged by setting parameter α equal to zero. Hence, technological competitiveness is only determined by the intrinsic quality of a technology. Further, in this regime new technologies are highly compatible with old technologies. This situation is obtained by setting the intrinsic quality levels as follows: $Q_3 = 1$, $Q_2 = 0.5$, and $Q_1 = 0.25$.

The second technology adoption regime, labelled 'intermediate regime' is characterised by again perfect substitutability of quality and network size, but here new technologies are less compatible with old technologies than in the quality regime. The lower compatibility of succeeding technologies is obtained by decreasing the differences in their intrinsic quality levels: $Q_3 = 1$, $Q_2 = 0.75$, and $Q_1 = 0.5$. Thus, compared to the quality regime, the second best technology is more competitive *vis-à-vis* the best technology available.

The settings of the third adoption regime (the 'network' regime) are such that quality and network size are to some extent complementary, with $\alpha = 0.5$. Further, $Q_3 = 1$, $Q_2 = 0.75$, and $Q_1 = 0.5$. Under this regime, technological competitiveness is also determined by the market share of a technology. The other parameters remain constant across the three technology adoption regimes.[99]

Although the three adoption regimes we analyse here are not perfectly consistent with the cases described by Shy (1996), we can still base our expectations with regard to the outcomes of the simulations on the predictions of his model. According to Shy (1996), whenever new technologies are perfectly compatible with old technologies, the new technologies are adopted each period. Further, a decrease in the degree of compatibility between new and old technologies would increase the duration of each technology. Based on this, we may expect that the duration of a technology is higher in the intermediate regime than in the quality regime.

Further, Shy (1996) concludes that when newer technologies are not perfectly compatible with older technologies, new technologies are never adopted if consumers treat network size and technological advance as perfect complements, but may be adopted if they are treated as perfect substitutes. Therefore, we may expect from our simulations that duration will be highest under the network regime, although eventually newer technologies will be adopted, given that our parameters are not consistent with perfect complementarity between network size and quality.[100] Finally, we may expect to see that new technologies are not always adopted whenever they appear in both the quality and, more likely, in the intermediate regime. In both these regimes there is perfect substitutability between quality and network size and imperfect compatibility. However, Shy's result (of technologies being skipped occasionally under these conditions) very much relies on his notion of compatibility. Therefore, some scepticism with regard to this expectation is appropriate.

Figure 5.1a to 5.1c show the evolution of the market share of successive product technologies for the three adoption regimes, resulting from one

[99]The values of the other parameters are: $N(0) = 40$, $\rho_E = 0.15$, $\mu_\lambda = 1$, $\sigma_\lambda = 0.1$, $s_i = 250$, $\beta_1 = 500$, $Q_0 = 0$, $\theta = 0.99$, $\beta_2 = 0.6$, $\Phi_L = 0.6$, $\Phi_H = 1$, $\hat{s} = 25$, $\beta_3 = 10$, $\beta_4 = 0.01$, $\beta_5 = 0.0069$, $\sigma_\chi = 0.1$, $\beta_6 = 0.01$, $\rho_{im} = 20$, $\beta_7 = 0.1$, $\beta_8 = 0.1$, $\sigma_\eta = 0.5$, $\beta_9 = 0.05$, and $\beta_{10} = 0.8$. The results of the simulations are robust to small changes in the levels of all parameters.

[100]However, when α is set equal to one, the simulations of our model indeed show that new technologies are never adopted.

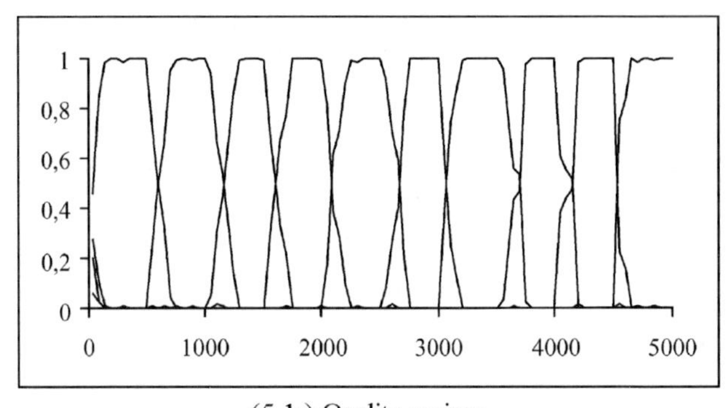

(5.1a) Quality regime

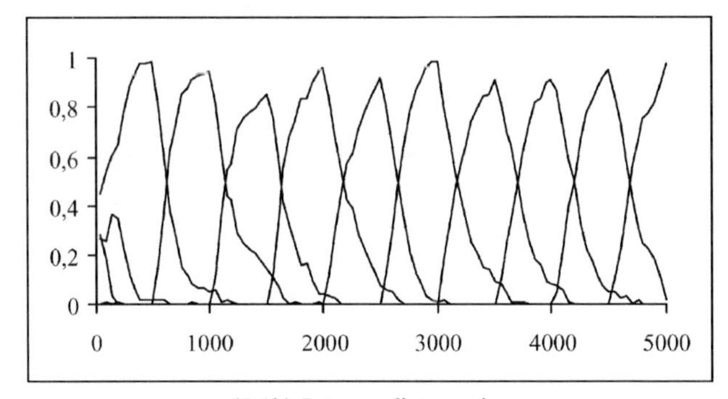

(5.1b) Intermediate regime

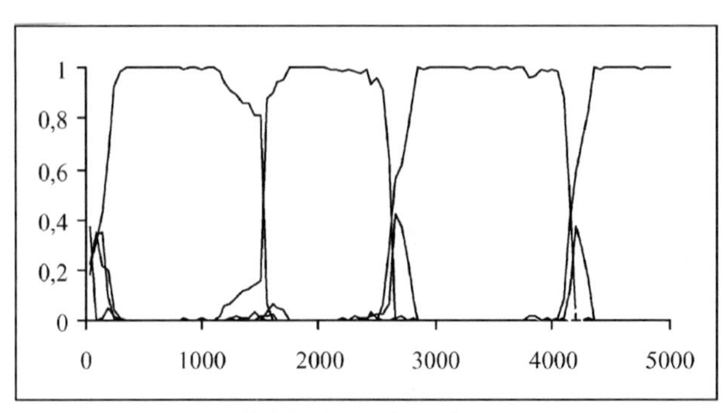

(5.1c) Network regime

Figure 5.1 Market Shares of Successive Product Technologies

simulation run of 5000 periods. As figure 5.1a shows, in the quality regime technologies have a rather short life. Whenever a new technology is introduced, it quickly replaces the previous technology and starts to dominate the market until again a new technology becomes available. In the intermediate regime, the replacement process is considerably slower, implying a longer duration of technologies (see figure 5.1b). Under this regime, none of the technologies persistently captures the entire market. The last figure shows that under the network regime, after a very short period of coexistence, one technology always dominates the market until it becomes obsolete[101], after which it is quickly replaced by a new technology that again dominates until its obsolescence. Thus, the outcomes of the simulations are fairly comparable to Shy's predictions. The duration of technologies is higher when compatibility between new and old technologies is lower. However, both under the quality regime as well under the intermediate regime, new technologies are always adopted.[102] Only under the network regime new technologies are never adopted until the existing dominant technology has become obsolete (i.e., until its intrinsic quality level Q_{Ψ} has dropped to Q_0).

How do the populations of firms evolve under these different regimes, and are there significant differences between the regimes? Figures 5.2 and 5.3 show the evolution of the number of firms and the concentration levels. Under the quality regime, we observe a gradually declining number of firms and increasing concentration levels. The intermediate regime exhibits a gradually growing population of firms with decreasing concentration levels. Finally, under the network regime we see on average a growing population of firms, however the growth rates fluctuate considerably. Further, new technology adoptions in the network regime are associated with a sharp decrease in the number of firms.

These figures, derived from one simulation run for each regime, indeed show that the regimes differ with regard to the population dynamics. But for a better assessment of the significance of these differences we calculated a number of statistics on the basis of the output of ten runs per regime[103], shown in table 5.2.

[101] I.e., until it becomes part of the class of old technologies.

[102] As mentioned before, this is due to the different notion of compatibility in Shy (1996).

[103] In order to keep the datasets at a reasonable size we have sampled each run only every 50 periods.

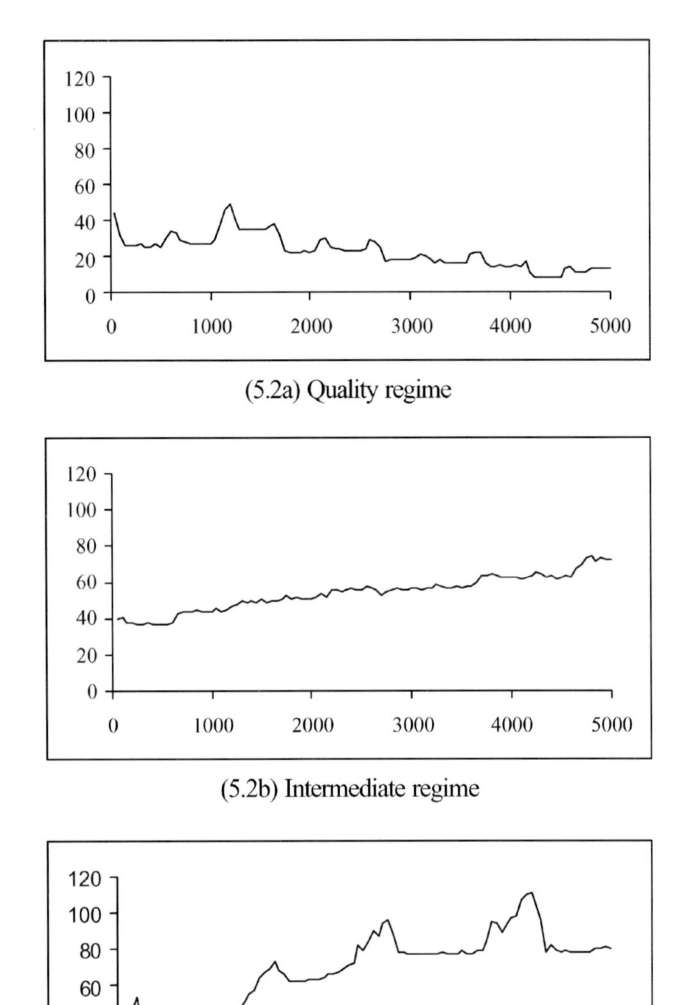

Figure 5.2 Number of Firms

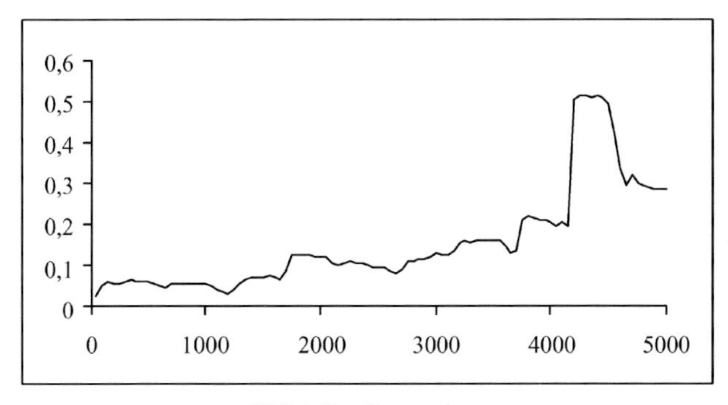

(5.3a) Quality regime

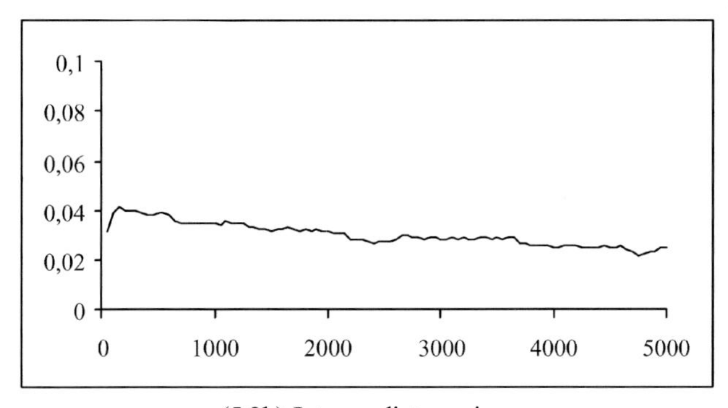

(5.3b) Intermediate regime

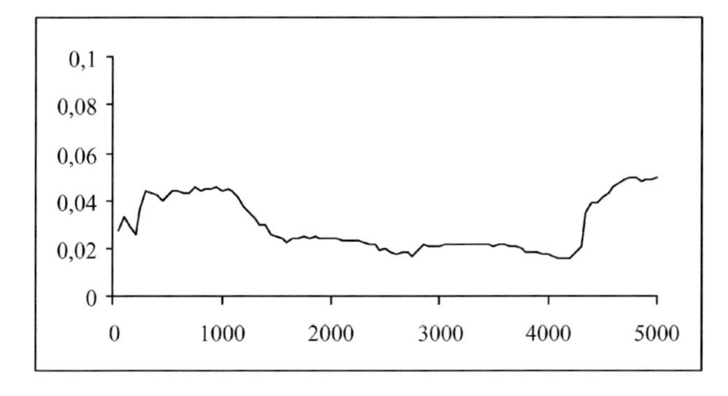

(5.3c) Network regime

Figure 5.3 Herfindahl Index

Table 5.2 Descriptive Statistics of Simulation Output

	Quality regime	Intermediate regime	Network regime
Number of firms	21.2 (1.27)	56.1 (2.34)	66.3 (1.70)
Herfindahl-index	12.5 (1.58)	3.57 (0.25)	3.44 (0.22)
Number of entrants over 50 periods	1.24 (0.08)	1.26 (0.06)	2.75 (0.05)
Survival rates (%)			
– short-run	43.9 (1.62)	52.1 (0.99)	49.8 (1.12)
– long-run	31.6 (1.74)	43.9 (0.78)	30.5 (0.63)
Mean age of			
– all exiters	913 (32.8)	1176 (53.3)	840 (41.3)
– all firms at t = 5000	1852 (177)	2937 (67.6)	2374 (118)

Note: Means over ten simulations. Standard errors in parentheses.

The first two rows of table 5.2 show the average number of firms and the average Herfindahl-index.[104] Next, we calculated the average number of entrants over 50 periods.[105] Entrants are defined as firms absent at the beginning of a 50-period era, but present at the end of the era (*vice versa* for exiters); incumbents are firms present throughout the sample period. Further, we calculated survival rates for entrants, both for the short-run (i.e., the proportion of entrants that survive at least 50 periods) as well as for the long run (proportion of entrants surviving at least 500 periods). Finally, we calculated the weighted[106] average age of all exiting firms at their year of exit, and the average age of all firms at $t = 5000$. Table 5.2 shows the averages of these variables per regime over all the sample periods of the ten simulations.

On average, the largest number of firms is found in the network regime, whereas the quality regime exhibits the smallest population. Not surprisingly then, the highest concentration levels are found under the quality regime, whereas the network regime produces the lowest concentration levels.

[104]The Herfindahl index is calculated as 100 times the sum of squared market shares.

[105]We did not calculate entry and exit rates, because they would not reflect real differences in, e.g., the ease of entry. If a regime is conducive to entrants the simulation results will show a higher number of firms in time than for a regime less conducive to entrants. Since the expected number of entrants in each period is fixed by the arrival rate ρ_E and identical across the three regimes, the entry conducive regime will show lower entry rates than the regime less conducive to entrants.

[106]For the weights we use the size of a firm (s_i in the model).

The highest number of entrants emerges under the network regime, which is approximately twice as high as under the other regimes.[107]

Concerning the survival rates, we see that the intermediate regime produces the highest probability for entrants to survive, both for the short run as well as the long run. The lowest short-run survival rates emerge under the quality regime. Long-run survival rates for the quality and the network regime are virtually similar.

The average age of exiting firms is highest under the intermediate regime, and lowest under the network regime. Further, at the end of the simulation the oldest population is found under the intermediate regime, whereas the youngest population is found under the quality regime.

All these results suggest that the three adoption regimes induce cross-sectional differences in the structural and dynamic properties of the population of firms. As such, our model indeed provides an additional explanation for the empirically observed sectoral variances. Although this result in itself is very important, we believe it is of equal interest that our model is also consistent with the stylised facts that empirical research in industrial economics has put forward so far. Among these "facts", the most essential for our model are: (*i*) persistence of market turbulence due to entry and exit, (*ii*) high infant mortality, negatively correlated with firm age, (*iii*) growth rates of firms that fall with age and with size, (*iv*) persistence of asymmetric performances, and (*v*) skewed and stable size distributions. We will therefore conclude this section by investigating whether our model is able to reproduce these stylised facts.

From the results that we have already mentioned in this section, it is obvious that the model reproduces the first stylised fact. Under all regimes, there is persistent market turbulence due to entry and exit. The second stylised fact (high infant mortality, negatively correlated with firm age) also emerges, except for the quality regime. Figure 5.4 shows the probability of exiting over the full simulation period given the age cohort[108] of an entrant for each regime.

Under the quality regime, infant mortality initially declines with age, however no entrant is able to live for more than approximately 4,500 periods. This leads to a rise in hazard rates for entrants with ages exceeding

[107]Note that we only sample every 50 periods. Therefore, despite the identical entry arrival rate of 0.15 per period, the three regimes produce different numbers of entrants due to variations in the number of firms that die before being observed in the sample. Hence, this number can be interpreted as a 'very short run' survival rate.

[108]Each age cohort covers 50 simulation periods.

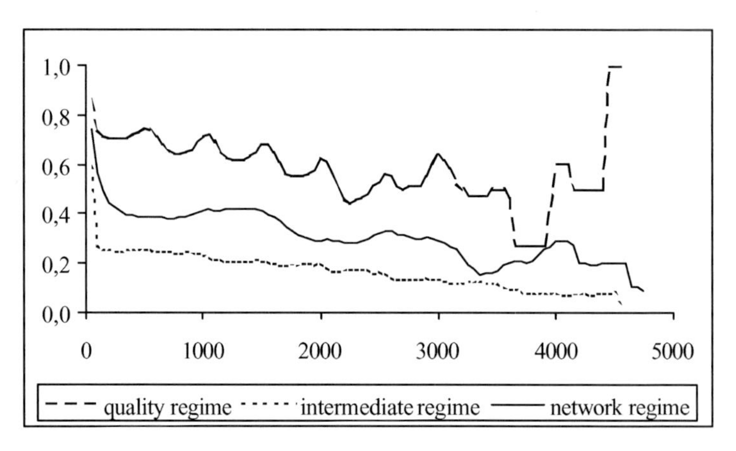

Figure 5.4 Hazard Rate as a Function of Age

4,000 periods. The other two regimes show very similar hazard rates that indeed decline as entrants mature. Hence, both the intermediate regime and the network regime reproduce this second stylised fact.

The third stylised fact (growth rates of firms that fall with age and with size) is of course imposed on the model by equation (5.7). Still, it might be interesting to learn about the emerging econometric regularities of the model. As in Dosi et al. (1995), we therefore run a number of regressions of the following form:

$$\ln\left(\frac{s_i(t+T)}{s_i(t)}\right) = q_0 + q_1 \ln s_i(t) + q_2 \ln a_i(t) + q_3(\ln s_i(t) \ln a_i(t)), \quad (5.12)$$

for $t = 50, 100, \ldots, 5000$ and $T = 50$ (regression 5.12a), and for $t = 500, 1000, \ldots, 5000$ and $T = 500$ (regression 5.12b). The results are listed in table 5.3.

Under all regimes this stylised fact is reproduced. All parameter estimates are significant at the 1 percent level, except for the ones indicated with an asterisk, which are only significant at the 10 percent level. Thus, both over the 50 period interval as well as over the 500 periods (except for the quality regime), initial size and initial age exert a negative impact on firm growth. Surprisingly, the interaction term exhibits in all cases a positive coefficient. Apparently the negative effect of, for instance, age on firm growth is attenuated for larger firms. Given the specification of equation (5.7), this emerging property is hard to explain. However, this regularity has been

Table 5.3 Regression Analyses of Firm Growth on Size and Age

	Constant	Size	Age	Size × Age	R^2-adjusted
Quality regime					
5.12a	0.595	−0.045	−0.085	0.006	0.098
5.12b	1.306	−0.135*	−0.202	0.020*	0.036
Intermediate regime					
5.12a	0.595	−0.055	−0.077	0.007	0.119
5.12b	1.128	−0.099	−0.151	0.012	0.170
Network regime					
5.12a	0.641	−0.058	−0.086	0.008	0.101
5.12b	1.707	−0.146	−0.235	0.020	0.204

observed before in empirical studies on firm growth. E.g., Evans (1987a, 1987b) found significant positive estimates for the variable indicating the interaction between age and size as well.

Persistence of asymmetric performances, the fourth stylised fact, also emerges from the model. To show this, we calculated for each 50th period the standard deviation of the mean of relative competitiveness of all firms (i.e., relative to the industry mean). These series are plotted in figure 5.5. As this figure shows, there is no convergence in the relative competitiveness of firms, hence in all regimes asymmetric performances are persistent. Especially under the quality regime the standard deviation clearly exhibits cyclical fluctuations, associated with the high speed of adoption of new technologies.

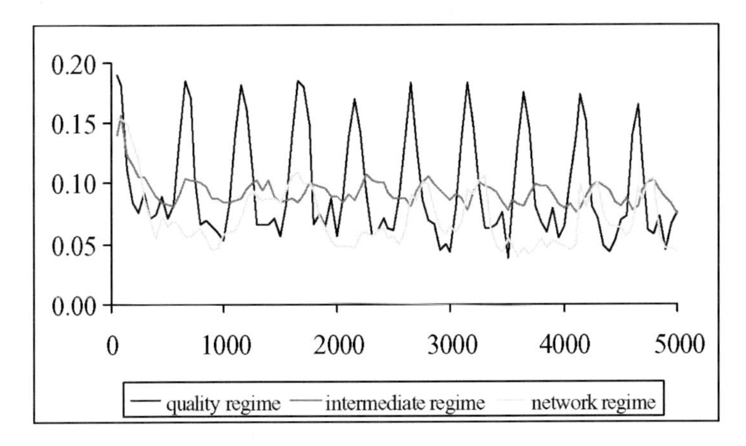

Figure 5.5 Standard Deviation of Average Relative Competitiveness

The evidence for the last stylised fact mentioned in section two, regarding the skewed and stable size distributions, is shown by figure 5.6a to 5.6c. These figures plot the log of firm sizes $s_i(t)$ on the vertical axis and the log of the firms' ranks $rank_i(t)$ on the horizontal axis (firms are ranked according to their size, in descending order) for $t = 500, 1000, \ldots, 5000$ for the same simulation run underlying figure 5.1.

All three regimes produce skew firm size distributions, but there are some interesting differences. The quality regime seems to produce the most skewed distribution, whereas the least stable size distribution emerges under the network regime. These findings, derived from visual inspection, are corroborated by the results from running the following regression:

$$\ln s_i(t) = A + B \ln rank_i(t), \tag{5.13}$$

for $t = 250, 750, \ldots, 4750$ (regression 5.13a), and for $t = 500, 1000, \ldots, 5000$ (regression 5.13b) over all simulation runs. Table 5.4 shows the results (all parameter estimates are significant at the 1 percent level).

Indeed, the quality regime shows the highest absolute value for B, and hence produces the most skewed size distribution. Further, we observe that it hardly matters when the regressions are executed (i.e., either at the end of a depreciation period or in the middle of it), indicating the stability of the size distribution.

In conclusion, the simulations of the three technology adoption regimes show significant cross-sectional differences with regard to the static and dynamic properties of the firm population. The quality regime produces the smallest and youngest population of firms, whereas long run survival seems to be easiest under the intermediate regime. Still, because of a higher number of entrants, the largest firm population emerges under the network regime. Hence, these results strongly suggest a relationship between the timing and frequency of new technology adoptions and the dynamics of the firm population. Furthermore, the model is consistent with a number of stylised facts observed in the empirical literature.

5.5 INTERACTION BETWEEN ADOPTION REGIMES AND TECHNOLOGICAL REGIMES

As mentioned in the introduction, one of the aims of our model was to introduce demand side considerations as an alternative to more supply side oriented explanations for industrial demography, such as the concept of

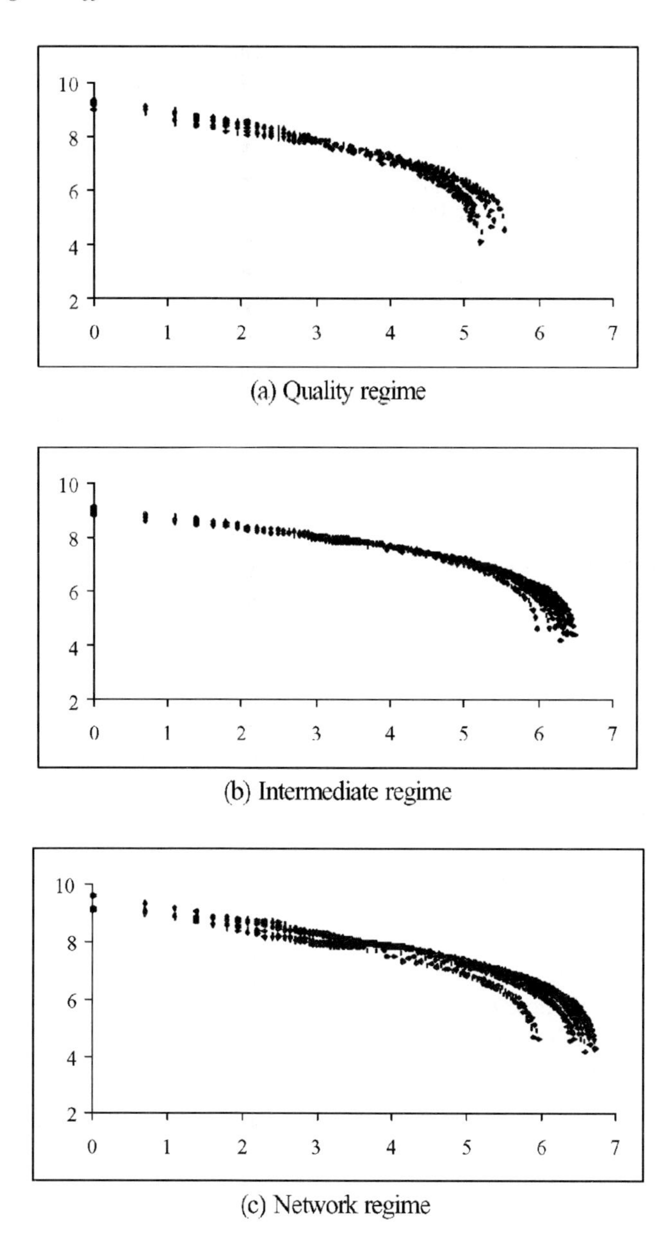

Figure 5.6 Size Distributions

Table 5.4 Regression Analyses of Firm Size on Rank

	Constant	Ln $rank_i$	Adjusted R^2
Quality regime			
5.13a	10.4	−0.93	0.84
5.13b	10.5	−0.90	0.83
Intermediate regime			
5.13a	10.7	−0.80	0.79
5.13b	10.8	−0.81	0.80
Network regime			
5.13a	10.9	−0.80	0.75
5.13b	11.0	−0.82	0.77

technological regimes. By varying a number of parameters determining the technological competitiveness of firms, we have designed three technology adoption regimes and investigated the differences in the firm population between these regimes. However, to some extent our model allows also for investigating the impact of different technological regimes on the industrial demography, and the interaction between adoption regimes and technological regimes. The present section will deal with these issues. By selecting and varying a number of parameters reflecting the level of cumulativeness and spillovers[109] of technological knowledge, we will run and compare several simulations of different technological regimes under the three technology adoption regimes.

Chapter 3 has shown that the conditions determining the technological regime have a strong impact on the patterns of innovative activities of an industry, as well as on the ease and the impact of imitation. Since in our model innovation is exogenous, we cannot fully simulate the different conditions underlying technological regimes. Only to the extent that these conditions apply to ease and impact of imitation we can analyse the effect of different technological regimes in our model.

To investigate whether our model is still able to produce the regularities predicted by the technological regime framework, we first have to identify the parameters reflecting the cumulativeness and spillover conditions. For cumulativeness we have to find the variable that indicates best to what

[109]For simplicity, we will reduce the number of conditions determining a technological regime to only two: cumulativeness and spillover effects. The latter captures both the appropriability and knowledge conditions from the original technological regime framework. Hence, differences in the opportunity levels regarding innovations are not considered here.

extent the acquisition of knowledge is a cumulative process. We propose to vary the parameter governing the 'penalty' rates for continuing firms (β_{10}). This penalty rate can be interpreted as a measure indicating to what extent the knowledge and experiences of a firm with its existing product technologies carry over to the new technology it adopts. For instance, if β_{10} is equal to β_2 (the penalty rate for entrants), the subfirm adopting the new technology starts at a competitiveness level equal to that of an entrant.[110] In this case the subfirm has no direct advantage over entrants with regard to the new technology (the accumulated experience with its existing technologies does not carry over to the subfirm with the new technology). However, the higher β_{10} (relative to β_2), the more the knowledge of new technology is based on the knowledge of previous technologies, and hence the more a new subfirm benefits from its accumulated experience *vis-à-vis* new firms entering with the same technology.[111]

For varying the spillover effects, it seems natural to vary the parameter determining the probability to receive an imitation draw (i.e., ρ_{im}). Since we have defined the process of imitating as acquiring the knowledge of applying a superior technology, this parameter reflects the ease of knowledge of new product technologies to flow to imitators. However, this parameter only reflects spillovers between continuing firms, not from continuing firms to entrants. Of course, in a strict sense, entrants are considered as continuing firms in the model from the moment they have entered. Thus, when the arrival rate of imitations is higher, the very young firms have a higher probability to receive an imitation draw as well. On the other hand, since more continuing firms will imitate, the average competitiveness level will be higher, which decreases the probability for entrants to survive. Several simulation runs with different imitation arrival rates indeed show that this last effect dominates: higher levels of spillovers generally lead to less (successful) entry, *ceteris paribus*. Since this inconsistency is due to the fact that entrants do not directly benefit from higher spillover levels, we consider it appropriate to increase the arrival rate of entrants (ρ_{ent}) too when spillovers levels

[110]I.e., relative to the firms' genotype organisational competitiveness levels λ_i.

[111]Of course, it may happen occasionally that an entrant with a certain technology receives an imitation draw quickly after it has entered. If β_{10} is high, this entrant will also benefit from its experience with the inferior technology, despite the short time it has been employing it. This will not seriously effect the results however, because the overall competitiveness of this entrant is at least for some time also determined by the competitiveness of the subfirm employing the inferior technology for which the entry penalty rate β_2 is still effective. Besides that, this firm already employed the initial technology successfully and therefore must possess some crucial knowledge about it.

are higher. In that case, more entrants have access to the knowledge necessary to imitate an existing technology. The entering firms will then also benefit directly from high spillover conditions.

In the next experiments we will consider three levels of cumulativeness and spillover conditions (low, medium, high) for each of the technology adoption regimes of the previous section. The specific parameter settings are as follows. Low cumulativeness means no direct advantage for subfirms relative to entrants ($\beta_{10}=\beta_2=0.6$). Medium levels correspond to the parameter settings of the previous section ($\beta_{10}=0.8$), whereas high cumulativeness levels are set by $\beta_{10}=1$. With regard to spillover effects, low levels of spillovers are set by $\rho_{\mathrm{I}}=10$ and $\rho_{ent}=0.1$. Medium levels again correspond to the parameter settings of the previous section (hence, $\rho_{\mathrm{I}}=20$ and $\rho_{ent}=0.15$). High spillover conditions are set by $\rho_{\mathrm{I}}=30$ and $\rho_{ent}=0.2$.

Hence, for each adoption regime we will consider nine different combinations of cumulativeness and spillover conditions. To get a general impression of the effect of varying the cumulativeness and spillover conditions, we will show the means and their standard errors of the same variables listed in table 5.2 under these various conditions over ten simulation runs (except for the number of entrants).[112] For an assessment of the significance of the effects of different cumulativeness and spillover conditions, and to investigate the extent to which these conditions interact, we will additionally perform the following regression analyses. Each of the six variables will be regressed on a number of dummy variables, indicating the various cumulativeness and spillover conditions and the potential interaction between them. More specific, for each adoption regime we will estimate the following equation:

$$\begin{aligned} Y = \beta_1 + \beta_2 Cm\_l + \beta_3 Cm\_h + \beta_4 Sp\_l + \beta_5 Sp\_h + \beta_6 (Cm\_lxSp\_l) \\ + \beta_7 (Cm\_lxSp\_h) + \beta_8 (Cm\_hxSp\_l) + \beta_9 (Cm\_hxSp\_h) + \varepsilon, \end{aligned} \tag{5.14}$$

where Y is the dependent variable under consideration, Cm_l equals one if cumulativeness conditions are low and zero otherwise, Cm_h equals one if cumulativeness is high and zero otherwise, Sp_l equals one if spillover conditions are low and zero otherwise, and Sp_h equals one if spillovers are high and zero otherwise. From this configuration of dummy variables it follows that the technological regime with medium cumulativeness and

[112]We will not display the evolution of market shares of succeeding technologies for the different technological regime conditions, because these patterns hardly show any differences *within* each technology adoption regime.

spillover conditions will be the reference regime, implying that the estimates for β_1 will be equal to the means of the variables listed in table 5.2.

Based on the technological regime framework, we would expect that, *ceteris paribus*, higher cumulativeness conditions generate a lower number of firms and higher concentration levels, lower survival rates and relatively young exiting firms, leading ultimately to an older population of firms. With regard to spillover conditions we would generally expect opposite regularities to emerge from the simulation. We have no clear expectations regarding the signs of the interaction effects. As mentioned, in the literature on technological regimes often a distinction is made between two regimes (Schumpeter I vs. II) with opposite spillover and cumulativeness conditions. However, no explicit reference is made regarding the interaction between them. By impeding the persistence of monopolistic advantages, high spillover conditions hinder innovative firms to become large, whereas low cumulativeness conditions facilitates innovative entry (compared to high cumulativeness conditions). Therefore, both high spillovers and low cumulativeness *independently* impose restrains on concentration levels. To what extent these conditions reinforce each other is unclear, however. Studying the significance of the interaction between the spillover and cumulativeness dummies may partly illuminate this issue.

As expected, in all three technology adoption regimes we see that, *ceteris paribus*, higher spillover conditions lead to a higher number of firms and to lower concentration levels. Further, with regard to cumulativeness conditions, we see that in general the number of firms decreases and the concentration increases with higher cumulativeness. Only under the network regime there is no clear relationship between cumulativeness on the one hand and the number of firms and concentration levels on the other.

Table 5.5 Number of Firms

	Quality regime				Intermediate regime				Network regime		
	Sp_l	*Sp_m*	*Sp_h*		*Sp_l*	*Sp_m*	*Sp_h*		*Sp_l*	*Sp_m*	*Sp_h*
Cm_l	22.4	34.4	48.6	*Cm_l*	51.2	67.6	83.2	*Cm_l*	46.5	63.0	87.8
	(1.20)	(1.05)	(1.53)		(1.50)	(3.17)	(3.38)		(2.68)	(1.53)	(2.89)
Cm_m	17.2	21.2	30.8	*Cm_m*	40.4	56.1	72.1	*Cm_m*	48.5	66.3	84.4
	(0.70)	(1.27)	(1.15)		(2.04)	(2.34)	(2.39)		(2.24)	(1.70)	(3.91)
Cm_h	10.2	12.1	16.0	*Cm_h*	31.1	36.8	42.6	*Cm_h*	39.9	49.2	71.3
	(0.88)	(1.33)	(0.74)		(1.29)	(1.36)	(2.09)		(2.79)	(1.74)	(3.78)

Note: Means over ten simulations. Standard errors in parentheses.

Table 5.6 Herfindahl-index

	Quality regime				Intermediate regime				Network regime		
	Sp_l	*Sp_m*	*Sp_h*		*Sp_l*	*Sp_m*	*Sp_h*		*Sp_l*	*Sp_m*	*Sp_h*
Cm_l	10.1	6.13	4.64	*Cm_l*	4.50	3.62	2.97	*Cm_l*	5.37	3.53	2.77
	(0.79)	(0.44)	(0.37)		(0.44)	(0.25)	(0.12)		(0.54)	(0.17)	(0.28)
Cm_m	12.6	12.5	7.70	*Cm_m*	5.23	3.57	2.90	*Cm_m*	4.89	3.44	2.99
	(0.97)	(1.58)	(1.15)		(0.48)	(0.25)	(0.26)		(0.41)	(0.22)	(0.20)
Cm_h	31.2	23.4	14.9	*Cm_h*	6.37	5.29	4.89	*Cm_h*	5.72	4.89	3.39
	(5.61)	(3.38)	(1.58)		(0.28)	(0.38)	(0.37)		(0.67)	(0.39)	(0.26)

Note: Means over ten simulations. Standard errors in parentheses.

The regression analyses, presented in table 5.7, show that with regard to the number of firms the differences due to variations in the technological regime parameters are significant. Only under the network regime the dummy for lower cumulativeness is not statistically significant. Interesting differences emerge with regard to the interaction effects. Under the quality regime, the combination of low cumulativeness and low spillovers significantly decreases the number of firms. Thus the positive effect of lower cumulativeness on the number of firms is almost completely offset by the negative effect of lower spillovers under this regime. In the opposite case, the negative effect of higher cumulativeness is again offset by the positive effect of high spillovers, but here the former effect dominates.

In cases with opposite spillover and cumulativeness conditions, we observe that the dummy for low cumulativeness and high spillover is significantly positive under the quality regime: low cumulativeness and high spillovers reinforce each other in this case. However, no significant interaction emerges in the opposite scenario, i.e., no significant interaction is observed between high cumulativeness and low spillovers under the quality regime. Under the intermediate regime, this latter interaction is significantly positive. Here, high spillovers and low cumulativeness *together* result in a higher number of firms than would be expected on the basis of these two effects individually. The other significant interaction under the intermediate regime emerges with high spillover and high cumulativeness. As under the quality regime, the negative effect of higher cumulativeness is again offset by the positive effect of high spillovers, where the former effect dominates. Finally, under the network regime no significant interaction is observed between cumulativeness and spillovers.

Table 5.7 Regression Analyses of Number of Firms and Herfindahl-index on Technological Regime Conditions

	Number of firms			Herfindahl-index		
	Quality regime	Intermediate regime	Network regime	Quality regime	Intermediate regime	Network regime
Constant	21.2**	56.1**	66.3**	12.5**	3.57**	3.44**
	(18.8)	(24.6)	(24.4)	(5.32)	(10.8)	(8.96)
Cm_l	13.2**	11.5**	−3.36	−6.42	0.05	0.09
	(8.32)	(3.58)	(−0.88)	(−1.93)	(0.11)	(0.17)
Cm_h	−9.02**	−19.3**	−17.1**	10.9**	1.72**	1.44**
	(−5.67)	(−5.96)	(−4.46)	(3.26)	(3.66)	(2.66)
Sp_l	−4.02**	−15.7**	−17.8**	0.06	1.66**	1.45**
	(−2.53)	(−4.86)	(−4.65)	(0.02)	(3.54)	(2.66)
Sp_h	9.60**	16.0**	18.1**	−4.87	−0.68	−0.45
	(6.03)	(4.96)	(4.72)	(−1.46)	(−1.44)	(−0.83)
Cm_l x Sp_l	−8.01**	−0.70	1.34	3.92	0.78	0.39
	(−3.56)	(−0.15)	(0.25)	(0.83)	(−1.18)	(0.51)
Cm_l x Sp_h	4.63*	−0.48	6.70	3.38	0.02	−0.31
	(2.06)	(−0.10)	(1.23)	(0.72)	(0.04)	(−0.41)
Cm_h x Sp_l	2.03	10.0*	8.55	7.73	−0.57	−0.61
	(0.90)	(2.19)	(1.58)	(1.64)	(−0.87)	(−0.80)
Cm_h x Sp_h	−5.72**	−10.3*	4.00	−3.71	0.28	−1.05
	(−2.54)	(−2.25)	(0.74)	(−0.79)	(0.42)	(−1.36)
Adjusted R^2	0.91	0.84	0.78	0.53	0.51	0.39

Note: *t*-values are in parentheses.
*Significant at the 5%-level.
**Significant at the 1%-level.

With regard to concentration levels the dummies for lower cumulativeness and higher spillovers are insignificant across all adoption regimes. Further, the dummy for low spillovers is insignificant under the quality regime, but significant under the other regimes. Finally, none of the dummies covering the interaction effects is statistically significant under any of the three adoption regimes. Thus, concentration levels are less affected by varying the cumulativeness and spillover conditions than the total number of firms, indicating that most of the differences appear in the lower firm size classes.

With respect to short run survival rates, the relationship between cumulativeness and this variable is as expected, albeit rather weak under the network regime: higher cumulativeness is generally associated with lower short run survival. However, higher spillover levels lead to *lower* short run

Table 5.8 Short Run Survival Rates

	Quality regime				Intermediate regime				Network regime		
	Sp_l	*Sp_m*	*Sp_h*		*Sp_l*	*Sp_m*	*Sp_h*		*Sp_l*	*Sp_m*	*Sp_h*
Cm_l	50.1	49.6	46.5	*Cm_l*	62.6	56.0	52.1	*Cm_l*	54.1	49.6	49.4
	(1.46)	(1.39)	(1.70)		(0.77)	(1.34)	(1.83)		(1.70)	(0.81)	(0.97)
Cm_m	48.5	43.9	42.0	*Cm_m*	61.6	52.1	47.8	*Cm_m*	54.9	49.8	49.6
	(1.54)	(1.62)	(1.22)		(2.37)	(0.99)	(1.58)		(1.81)	(1.12)	(0.60)
Cm_h	36.6	30.7	30.0	*Cm_h*	52.0	34.5	28.2	*Cm_h*	50.4	45.4	46.9
	(3.01)	(4.54)	(0.95)		(2.02)	(1.96)	(2.45)		(1.54)	(1.91)	(1.21)

Note: Means over ten simulations. Standard errors in parentheses.

survival rates (again a rather weak effect under the network regime). Still, this is not surprising, since high spillovers increase the industrial average competitiveness, making it harder for entrants to survive.

As table 5.9 shows, the regression analyses generally confirm this. Compared to the technological regime with medium cumulativeness and spillover levels, higher cumulativeness significantly decrease the short run survival rates. Further, lower spillovers significantly increase the short run survival, except under the quality regime. The dummies for low cumulativeness and high spillovers are all statistically insignificant. Finally, only one interaction dummy is significant. Under the intermediate regime, the negative effect of high cumulativeness on short run survival is almost completely offset by the (unexpected) positive effect of low spillovers.

For long run survival rates, the picture is rather similar with regard to cumulativeness conditions. Under the quality and the intermediate regime, higher cumulativeness decreases long run survival, whereas long run survival under the network regime is not effected by different cumulativeness conditions. Only under the intermediate regime higher spillover conditions seem to have a (negative) effect on long run survival.

Again, these observations are confirmed by the regression analyses (see table 5.9). Under the network regime, none of the dummies is statistically significant, whereas under the quality regime only the dummy for high cumulativeness is significant. Further, under the intermediate regime the dummies for low spillovers (positive) and high cumulativeness (positive) show significant estimates. Finally, all interaction effects are statistically insignificant under all three adoption regimes.

For the mean age of exiting firms, the picture is quite diverse. Under the quality regime, higher spillovers tend to increase the average age of exiters

Table 5.9 Regression Analyses of Survival Rates on Technological Regime Conditions

	Short run survival rate			Long run survival rates		
	Quality regime	Intermediate regime	Network regime	Quality regime	Intermediate regime	Network regime
Constant	43.9**	52.1**	49.8**	31.6**	43.9**	30.5**
	(19.9)	(29.1)	(36.4)	(17.2)	(25.2)	(21.9)
Cm_l	5.70	3.89	−0.23	1.63	0.93	−1.12
	(1.82)	(1.54)	(−0.88)	(0.63)	(0.11)	(−0.57)
Cm_h	−13.2**	−17.5**	−4.38*	−8.38**	−13.0**	−2.80
	(−4.22)	(−6.94)	(2.26)	(−3.22)	(5.28)	(−1.42)
Sp_l	4.57	9.52**	5.06**	−1.98	6.17**	0.08
	(1.46)	(3.77)	(2.62)	(−0.76)	(2.51)	(0.04)
Sp_h	−1.87	−4.30	−0.24	−2.26	−2.57	0.78
	(−0.60)	(−1.70)	(−0.12)	(−0.87)	(−1.05)	(0.40)
Cm_l x Sp_l	−4.11	−2.91	−0.54	−0.23	−4.59	0.14
	(−0.93)	(−0.81)	(−0.20)	(−0.06)	(−1.32)	(0.05)
Cm_l x Sp_h	−1.22	0.47	0.10	3.21	2.03	0.80
	(−0.28)	(0.13)	(0.04)	(0.87)	(0.58)	(0.29)
Cm_h x Sp_l	1.29	7.91*	−0.11	3.75	5.32	1.38
	(0.29)	(2.21)	(−0.04)	(1.64)	(1.53)	(0.49)
Cm_h x Sp_h	1.14	−2.00	1.74	2.66	−4.69	1.70
	(0.26)	(−0.56)	(0.64)	(0.72)	(−1.35)	(0.61)
Adjusted R^2	0.51	0.78	0.25	0.26	0.66	0.00

Note: *t*-values are in parentheses.
*Significant at the 5%-level.
**Significant at the 1%-level.

Table 5.10 Long Run Survival Rates

	Quality regime				Intermediate regime				Network regime		
	Sp_l	*Sp_m*	*Sp_h*		*Sp_l*	*Sp_m*	*Sp_h*		*Sp_l*	*Sp_m*	*Sp_h*
Cm_l	31.0	33.2	34.1	*Cm_l*	46.4	44.8	44.3	*Cm_l*	29.6	29.4	31.0
	(1.11)	(1.31)	(1.22)		(1.30)	(1.77)	(1.83)		(1.32)	(0.98)	(1.06)
Cm_m	29.6	31.6	29.3	*Cm_m*	50.1	43.9	41.3	*Cm_m*	30.6	30.5	31.3
	(1.41)	(1.74)	(1.63)		(2.02)	(0.78)	(1.55)		(2.06)	(0.63)	(1.30)
Cm_h	25.0	23.2	23.6	*Cm_h*	42.4	30.9	23.7	*Cm_h*	29.2	27.7	30.2
	(1.77)	(3.60)	(1.50)		(1.63)	(2.20)	(2.12)		(1.73)	(1.48)	(1.48)

Note: Means over ten simulations. Standard errors in parentheses.

(as expected), whereas under the intermediate regime the opposite tendency is observed. No clear relationship between spillover levels and the mean age of exiters emerges under the network regime. With regard to cumulativeness, only the regularities under the network regime match our expectations. Here, higher cumulativeness decreases the average age of exiters. This relationship seems to be reversed under the intermediate regime, whereas no clear effects of cumulativeness conditions are observed under the quality regime.

Naturally, the regression statistics in table 5.12 show a dispersed picture as well, most notably for the quality regime. Both the dummies for low and high cumulativeness are statistically significant, with both being negative. The dummy for low spillovers is also significant under this regime, with the expected negative sign. Under the intermediate regime, both the dummies for low cumulativeness and high spillover are significant, but both have a negative sign where we would expect positive ones. Finally, under the network regime the dummy for high cumulativeness and the dummy for low spillovers are significant, with the expected negative sign.

Regarding the interaction effects, it is interesting to see that only under the quality regime significant interactions emerge, where the combined effects seem to be less strong than the sum of individual effects. This applies to the combinations of low spillovers and low cumulativeness, low spillovers and high cumulativeness, and high spillovers and high cumulativeness.

The average age of the total population at the ultimate period (table 5.13) generally increases with cumulativeness and spillover conditions under the intermediate regime and, to a lesser extent, under the quality regime, whereas no relationships emerge under the network regime. The regression

Table 5.11 Mean Age of Exiters

	Quality regime				Intermediate regime				Network regime		
	Sp_l	*Sp_m*	*Sp_h*		*Sp_l*	*Sp_m*	*Sp_h*		*Sp_l*	*Sp_m*	*Sp_h*
Cm_l	676	759	834	*Cm_l*	1039	819	444	*Cm_l*	874	883	814
	(21.2)	(17.8)	(23.3)		(37.2)	(28.1)	(24.3)		(42.2)	(25.8)	(39.6)
Cm_m	666	913	890	*Cm_m*	1297	1176	682	*Cm_m*	759	840	755
	(25.7)	(32.8)	(35.6)		(43.4)	(53.3)	(75.6)		(32.3)	(41.3)	(35.2)
Cm_h	659	745	861	*Cm_h*	1224	1232	612	*Cm_h*	699	653	559
	(37.0)	(51.3)	(27.1)		(45.8)	(58.5)	(84.1)		(34.2)	(24.2)	(34.9)

Note: Means over ten simulations. Standard errors in parentheses.

Table 5.12 Regression Analyses of Mean Age of Exiters and Mean Firm Age at $t = 5000$ on Technological Regime Conditions

	Mean age of exiters			Mean age at $t = 5000$		
	Quality regime	Intermediate regime	Network regime	Quality regime	Intermediate regime	Network regime
Constant	913**	1176**	840**	1852**	2937**	2374**
	(28.8)	(22.0)	(24.0)	(8.30)	(31.9)	(22.0)
Cm_l	−154**	−357**	43.3	−409	−909**	−141
	(−3.44)	(−4.72)	(0.88)	(−1.30)	(−6.98)	(−0.92)
Cm_h	−168**	55.7	−187**	865**	890**	64.2
	(−3.75)	(0.74)	(−3.78)	(2.74)	(6.83)	(0.42)
Sp_l	−247**	121	−80.8**	−571	−662**	−121
	(−5.51)	(1.60)	(−1.64)	(−1.81)	(−5.08)	(−0.79)
Sp_h	−22.8	−495**	−85.1	172	357**	148
	(−0.51)	(−6.54)	(−1.72)	(0.55)	(2.74)	(0.97)
Cm_l x Sp_l	164**	98.2	71.7	426	572**	−58.2
	(2.58)	(0.92)	(1.03)	(0.95)	(3.10)	(−0.27)
Cm_l x Sp_h	98.4	119	16.2	65.9	−69.1	−22.6
	(1.55)	(1.11)	(0.23)	(0.15)	(−0.38)	(−0.11)
Cm_h x Sp_l	161**	−129	127	−33.9	−397*	−251
	(2.54)	(−1.21)	(1.81)	(−0.08)	(−2.16)	(−1.16)
Cm_h x Sp_h	139*	−125	−8.83	−524	−86.2	−202
	(2.19)	(−1.17)	(−0.13)	(−1.17)	(−0.47)	(−0.94)
Adjusted R^2	0.44	0.75	0.43	0.26	0.86	0.10

Note: *t*-values are in parentheses.
*Significant at the 5%-level.
**Significant at the 1%-level.

Table 5.13 Mean Age at $t = 5000$

	Quality regime				Intermediate regime				Network regime		
	Sp_l	*Sp_m*	*Sp_h*		*Sp_l*	*Sp_m*	*Sp_h*		*Sp_l*	*Sp_m*	*Sp_h*
Cm_l	1298	1444	1682	*Cm_l*	1938	2029	2316	*Cm_l*	2055	2234	2359
	(82.3)	(81.5)	(75.5)		(59.8)	(58.4)	(75.6)		(109)	(68.8)	(85.3)
Cm_m	1281	1852	2024	*Cm_m*	2275	2937	3294	*Cm_m*	2253	2374	2522
	(122)	(177)	(132)		(37.3)	(67.6)	(47.3)		(174)	(118)	(56.4)
Cm_h	2112	2717	2365	*Cm_h*	2768	3827	4098	*Cm_h*	2067	2438	2385
	(424)	(390)	(183)		(157)	(129)	(120)		(72.9)	(122)	(116)

Note: Means over ten simulations. Standard errors in parentheses.

statistics in table 5.12 also show no significant dummies for the network regime. For the quality regime, only the dummy for high cumulativeness is statistically significant. The dummies for cumulativeness have the expected sign and are both significant under the intermediate regime. The dummies for spillover conditions are also statistically significant, however their signs do not match the expectations based on the technological regime framework.

Interaction effects are only significant under the intermediate regime this time. The (expected) negative effect of low cumulativeness combined with the (unexpected) negative effect of low spillovers results in an older population than we would expect from those two effects individually. The positive effect of high cumulativeness on the average age of the total population is more than offset by the negative effect of low spillovers when these are combined.

In conclusion, we observe that with regard to the cumulativeness conditions the regularities predicted by the technological regime framework are reproduced by the model. In general, we see a smaller, more concentrated and eventually older population of firms when cumulativeness conditions are high. Spillover conditions are in line with the expectations regarding the number of firms and concentration levels. I.e., higher spillovers lead to a higher number of firms and to lower concentration levels. However, they do not increase survival rates or decrease the average age of the population.

A possible explanation for this is that the benefits of the incumbents from high spillovers are such that they easily imitate and survive, thereby increasing the industrial average competitiveness. This process leads both to an eventually older population, as well as to hard survival conditions for entrants. Another reason for this inconsistency with the technological regime framework is that our model does not allow for analysing the effect of the technological regime conditions on differences between the innovative activities of incumbents and entrants. Innovation is exogenous in our model, and perhaps endogenising the innovation process would make our model more consistent with the technological regime framework.

Finally, we observe that the regularities emerging under the network regime are the least affected by variations in spillover and cumulativeness conditions. The interaction effects between spillovers and cumulativeness are never significant under the network regime as well. Apparently, the emerging regularities under this regime are mainly determined by the network externalities among users and the resulting diffusion patterns of new product technologies. Under the quality and intermediate regime, we have found some significant interaction effects, however they do not appear to systematically affect the results.

5.6 CONCLUSIONS

This chapter has shown that differences in the timing and frequency of new product technology adoptions provide an additional explanation for sectoral variations in the dynamics of the firm population. Assuming that varying consumer preferences over technology advance and network size effects, and different degrees of compatibility between succeeding technologies explain why in some industries technologies are more often replaced than in others (see Shy, 1996), we analysed, by means of a simulation model, how the different replacement patterns would effect the dynamics of the firm population. We designed and investigated three different technology adoption regimes with the following underlying conditions: (*i*) a quality regime, in which quality and network size are regarded as perfect substitutes and new product technologies are highly compatible with old technologies, (*ii*) an intermediate regime, in which new product technologies are less compatible with old technologies, but where quality and network size are still regarded as perfect substitutes, and (*iii*) a network regime, in which network size and quality are regarded as complementary. By modelling the growth of a firm's competitiveness as a function of both the quality level and the market share of the product technologies it employs, and by tuning the parameters of this function to arrange the adoption regimes, the model produced the following results.

First of all, three rather different replacement patterns of technologies emerge. In the quality regime, technologies are continuously and rather quickly replaced by superior technologies as soon as they become available. In the intermediate regime, newer technologies still always replace older ones, but the duration of a technology is higher than under the quality regime. Finally, in the network regime eventually one technology always dominates the market until it becomes obsolete (despite the presence of superior technologies), after which it is quickly replaced by a new technology that again dominates until its obsolescence.

The second result is that the replacement patterns of product technologies clearly affect the dynamics of the firm population. The quality regime produces the smallest, but most dynamic population of firms, whereas the largest firm population emerges under the network regime. The intermediate regime exhibits the most stable population of firms, where long run survival is relatively easy.

The third result is that for all three regimes the model is able to reproduce a number of important stylised facts from industrial organisation. The model produces persistence of market turbulence due to entry and exit; high infant mortality, negatively correlated with firm age; growth rates of firms that fall with age and with size; persistence of asymmetric performances; and skewed and relatively stable size distributions.

The fourth result is that all these outcomes are obtained in the absence of replicator dynamics. There is no explicit relationship in our model between a firm's relative competitiveness and its growth rate. The only selection mechanism in our model is that a minimum level of relative competitiveness is required in order to survive. This rather simple mechanism turns out to be sufficient to produce meaningful results, consistent with the previously mentioned stylised facts.

The fifth and final result is derived from running the model under different technological regimes, represented by various cumulativeness and spillover conditions. In general, we see a smaller, more concentrated and eventually older population of firms when cumulativeness conditions are high. These regularities are in line with the technological regime framework. Spillover conditions, however, are only partly consistent with this framework. Higher spillovers indeed lead to a higher number of firms and lower concentration levels. However, they do not generally increase survival rates or decrease the average age of the population. The explanation for this is found in the trivial effect of high spillover conditions on the competitiveness of incumbent firms *vis-à-vis* entrants. High spillover conditions enable more continuing firms to imitate, which increases the industrial average competitiveness. This, in turn, deteriorates the general conditions for entrants and makes it more difficult for them to survive. Finally, we observed that the regularities emerging under the network regime are the least affected by varying cumulativeness and spillover conditions.

Endogenising the innovation process could make our model more consistent with the technological regime framework. Since innovation is exogenous in our model, we cannot analyse the effect of the technological regime conditions on differences between the innovative activities of incumbents and entrants. Perhaps future efforts in this direction will enable us to better assess the interaction between the demand side oriented adoption regimes and the more supply side oriented technological regimes.

6. CONCLUSIONS

This book has focussed on variances in the structural and dynamic properties of manufacturing industries. For many decades we have known empirically that industries differ widely in, for example, concentration levels, capital intensity, and entry and exit rates. And for many decades industrial economists, trying to find theoretical explanations for these observed differences, have mainly relied upon theories involving stable equilibrium solutions. Recently however, the increased access to firm-level databases has led to a number of robust empirical results that are hard to reconcile with the assumptions underlying the mainstream equilibrium theories and models of industrial economics, and with the results these models produce. Probably therefore we observe an increasing interest in theoretical alternatives based on more realistic assumptions regarding firm behaviour and the process of technological change. As such, these alternative theoretical frameworks may provide plausible explanations for the cross-sectional variations in the structures and dynamics of industries.

The major objective of this book was to find and test for explanations derived from these rather recently developed theoretical frameworks. By exploring the theoretical and empirical literature in industrial economics, by using a longitudinal firm-level database, and by developing and simulating a model on industrial dynamics, we obtained the following results. Empirically, we have shown that both the technological regime framework as well as the product or industry life cycle approach can explain differences between industries. Further, our model has shown that differences in the timing and frequency of new technology adoptions provide an additional explanation for the structural and dynamic differences between industries. In the next section we will briefly summarise the main chapters of this book. In the second and final section of this chapter we will suggest some directions for future research in industrial economics.

6.1 SUMMARY

Given the empirical nature of this book and the extensive use of the Statistics Netherlands manufacturing database throughout it, chapter 2 has illuminated three important issues related to empirical research in industrial economics. First of all, we have described in detail the firm-level data on which a major part of the research in this book is based. Second, by presenting a number of selected stylised facts we have aimed to highlight some robust results that have been established in industrial economics. Especially in the last fifteen years, the increased access to firm-level data has resulted in a collection of results that were found to be rather invariant across countries and over time. Finally, we have investigated whether these stylised facts could also be found in Dutch manufacturing.

Two important results emerged from this chapter. The first one was that the regularities emerging from the SN manufacturing database were to a large extent consistent with the stylised facts of industrial economics. Whether we looked at firm-level regularities, their aggregate impact on the manufacturing sector, or regularities at the industry level, practically all of them seemed to be consistent with earlier empirical findings in industrial economics. To mention just a few, we have found hazard rates of entering firms and growth rates of surviving entrants declining with age and (initial) size; persistence of productivity differentials between firms in the same industry; skewed but rather stable distributions of firm sizes; and finally, significantly positive correlations between entry and exit rates across industries.

The second result of chapter 2 concerns the observation that Dutch manufacturing industries widely differ in virtually all dimensions observable in the data. Although this result is far from new, we have argued that, despite the presence of a number of theoretical contributions that potentially can explain the observed differences, not much effort has been done in testing these theories with the increasingly available firm-level data. As such, this unexploited opportunity constituted one of the main objectives of this book: to investigate empirically which theoretical frameworks can explain the cross-sectional differences.

Based on the results of chapter 2, chapter 3 has selectively surveyed the theoretical literature in industrial economics. In our search for theoretical contributions that may explain the observed differences across industries in a dynamic and comprehensive way, we have selected two approaches and described them thoroughly in this chapter. We have argued that both the technological regime framework as well as the product life cycle approach are highly appropriate conceptual tools to study empirically the differences in cross-sectional regularities, as they both embody elements such as firm heterogeneity and technological uncertainty that are close to empirical substance. With regard to their potential power to explain differences in structures and dynamics between industries, we have argued the following.

The technological regime framework asserts that sectoral patterns of innovative activities are determined by the combination of opportunity, appropriability and cumulativeness conditions, and properties of the technological knowledge base underlying these activities. If entry, survival, growth and exit patterns are related to the rates and forms of organisation of innovations, then the observed inter-sectoral variety in structures and dynamics can be explained by differences in the underlying technological regimes. Alternatively, theories and models on product life cycles explain and depict the evolution of an industry's structural and dynamic properties over its lifetime. Based on this approach, the observed cross-sectional differences can be explained by the different evolutionary stages that industries occupy.

To what extent these theories could actually account for the cross-sectional variances was investigated in chapter 4. The first part of this chapter aimed to find empirical evidence for the technological regime framework. Based on this framework, we derived a number of hypotheses and tested these with the longitudinal firm-level data we had access to. This empirical analysis strongly suggested that differences in the structural (e.g., concentration levels) and dynamic (e.g., entry and exit rates) properties of industries are closely related to the set of opportunity, appropriability, cumulativeness and knowledge conditions underlying the innovative activities in an industry.

The second part of chapter 4 investigated whether the cross-sectional variations in structures and dynamics of industries could be explained by the different evolutionary stages the industries occupied. Based on a model by Klepper (1996), we again derived and tested a number of hypotheses. Although Klepper's model is primarily intended to depict the evolution of technologically progressive industries, the second part of chapter 4 has shown that applying the model to (basically mature) manufacturing industries in general, i.e., without any reference to their technological progressiveness, provides some interesting results. Industries in different evolutionary stages indeed show different regularities, which are partly in line with Klepper's model. Especially differences in the entry and survival rates of industries corroborate Klepper's hypothesis of a growing advantage of incumbents over entrants. Further, we have found evidence that a shakeout of firms is mainly caused by a reduction in entry rates, rather than an increase in exit rates. Again, this finding is consistent with Klepper's model. Apart from some minor differences, the regression analyses of the third and final part of chapter 4, which investigated the interaction between the two theoretical approaches, generally confirmed the results of the separate investigations.

In chapter 5 we have introduced a simulation model on industry dynamics. This model embodies three elements that, in our view, are not properly acknowledged by the technological regime framework and the product life cycle approach. First of all, models on product life cycles generally focus on the emergence and evolution of only one product and its associated technology. However, in many industries we observe that firms repeatedly introduce or adopt new product technologies that replace the older ones. Second, both these approaches do not explicitly consider differences in the technological properties of the goods produced by the industries. Finally, in models on technological regimes and on industry life cycles the growth of a firm is generally determined by its relative (technological) performance. However, empirical studies on firm growth do not provide much evidence supporting such a relationship. Most of these studies suggest that the size of a firm generally follows a random walk with a declining positive drift.

As mentioned, the model presented in chapter 5 includes these three elements. But the most interesting feature of this model is that it provides an additional explanation for sectoral variations in the structures and dynamics of the firm population in the following way. Assuming that varying consumer preferences over technology advance and network size effects, and different degrees of compatibility between succeeding technologies explain why in some industries technologies are more often replaced than in others (see Shy, 1996), our model showed how the different replacement patterns effect the structures and dynamics of the firm population.

More precisely, we designed and investigated three different technology adoption regimes with the following underlying conditions: (*i*) a quality regime, in which quality and network size are regarded as perfect substitutes and new product technologies are highly compatible with old technologies, (*ii*) an intermediate regime, in which new product technologies are less compatible with old technologies, but where quality and network size are still regarded as perfect substitutes, and (*iii*) a network regime, in which network size and quality are regarded as complementary. By modelling the growth of a firm's competitiveness as a function of both the quality level and the market share of the product technology it employs, and by tuning the parameters of this function to arrange the adoption regimes, the model produced the following results.

First of all, three rather different replacement patterns of technologies emerged. In the quality regime, technologies are continuously and rather quickly replaced by superior technologies as soon as they become available. In the intermediate regime, newer technologies still always replace older ones, but the duration of a technology is higher than under the quality regime. Finally, in the network regime eventually one technology always dominates the market until it becomes obsolete (despite the presence of superior technologies), after which it is quickly replaced by a new technology that again dominates until its obsolescence.

The second result was that the replacement patterns of product technologies clearly affected the dynamics of the firm population. The quality regime produced the smallest, but most dynamic population of firms, whereas the largest firm population emerged under the network regime. The intermediate regime exhibited the most stable population of firms, where long run survival is relatively easy. The third result was that for all three regimes the model is able to reproduce a number of important stylised facts from industrial organisation. The model produces persistence of market turbulence due to entry and exit; high infant mortality, negatively correlated with firm age; growth rates of firms that fall with age and with size; persistence of asymmetric performances; and skewed and relatively stable size distributions.

The fourth result was that all these outcomes were obtained in the absence of replicator dynamics. There is no explicit relationship in our model between a firm's relative competitiveness and its growth rate. The only selection mechanism in our model is that a minimum level of relative competitiveness is required in order to survive. This rather simple mechanism turned out to be sufficient to produce meaningful results, consistent with the previously mentioned stylised facts.

The fifth and final result was derived from running the model under different technological regimes, represented by various cumulativeness and spillover conditions. In general, we saw a smaller, more concentrated and eventually older population of firms when cumulativeness conditions are high. These regularities are in line with the technological regime framework. Spillover conditions, however, were only partly consistent with this framework. Higher spillovers indeed led to a higher number of firms and lower concentration levels. However, they did not generally increase survival rates or decreased the average age of the population. The explanation for this was found in the trivial effect of high spillover conditions on the competitiveness of incumbent firms *vis-à-vis* entrants. High spillover conditions enable more continuing firms to imitate, which increases the industrial average competitiveness. This, in turn, deteriorates the general conditions for entrants and makes it more difficult for them to survive. Finally, we observed that the regularities emerging under the network regime are the least affected by varying cumulativeness and spillover conditions.

In conclusion, this book has shown that theoretical approaches involving disequilibrium, bounded rationality and technological uncertainty provide plausible and empirically supported explanations for one of the main issues in industrial economics, namely the differences in structural and dynamic properties of industries.

6.2 SUGGESTIONS FOR FUTURE RESEARCH

In what ways can we further improve the empirical and theoretical foundations of industrial economics? Obviously, more data would be the most straightforward way to strengthen the empirical foundations. Although the longitudinal firm-level data at hand has already greatly enriched our knowledge about the firm-level dynamics and their impact at the level of the industry, still more information regarding investment behaviour, research and development, strategic alliances, et cetera, could further improve our understanding of industrial dynamics.

Regarding the empirical analyses of chapter 4, more data could certainly strengthen the results. For instance, variables indicating the actual opportunity, appropriability, cumulativeness and conditions of knowledge accumulation to determine the technological regime, or the actual age of an industry together with, e.g., the growth rate of patents to determine an industry's evolutionary stage, would undoubtedly provide a stronger basis for testing whether technological regimes and industry life cycles explain the observed differences between industries.

Concerning the theoretical foundations of industrial economics, we have argued in chapter 3 that the technological regime framework and the product life cycle approach are highly appropriate conceptual tools for studying industrial dynamics. Given their focus on the process of technological change and the co-evolving structural and dynamic properties of the industries, these two frameworks provide useful starting points in theorising on industrial economics. Advancing these theories therefore seems a fruitful and promising direction to be explored in industrial economics.

At present, such efforts are already being undertaken by scholars working on technological regimes. Marsili (1999), for instance, argues that the concept of appropriability conditions does not clearly distinguish differences in the ease with which internal and external firms can access a certain pool of technological opportunities. She introduces the notion of technological entry barriers in order to capture the dynamics of industrial competition driven by firms from outside the industry more accurately than the concept of appropriability, which applies to competitors from both inside and outside the industry.

The product or industry life cycle approach can also be fruitfully extended in our view. One could, for instance, include networks of exchange relationships that firms build up in time with their (financial) suppliers and customers. Besides dynamic increasing returns to technological change underlying present models, mutual learning and declining transaction costs associated with maintaining a network of exchange relationships may provide an additional growing competitive advantage of incumbent firms over entrants as the industry matures.

With regard to our model presented in chapter 5, we have already suggested one way of advancing our framework. Since innovation is exogenous in our model, we cannot analyse the effect of the technological regime conditions on differences between the innovative activities of incumbents and entrants. Hence, endogenising the innovation process could make our model more suitable for analysing the effects of various technological regime conditions. Furthermore, it may allow us to model technological entry barriers as well in order to take into account the suggestions made by Marsili (1999).

Another possibility to improve the behavioural assumptions of our model, and in fact of many more models on industry dynamics, is to further investigate the determinants of firm growth. As we have argued in chapter 5, econometric estimations regarding firm growth on the basis of relative firm performance (in terms of productivity or profitability) usually explain only a very small part of the total variance in growth rates (Geroski, 1998). For the outsider, the evolution of the size of a firm essentially follows a random walk. But what happens in the minds of the insiders? After all, the growth of a firm (in terms of capital and labour increases) is ultimately a managerial decision. But when exactly does the management of a firm decide to expand its capacity and, equally interesting, with how much will it increase (or decrease) the capacity of the firm? Attempts to identify the quantitative and qualitative variables that underlie the growth decisions of firms are very important as they may substantially improve the behavioural foundations of agent based models on industrial dynamics.

The final suggestion we make in this section regards the scope of industrial economics. Most theoretical and empirical work in industrial economics focuses on manufacturing. However, in most industrialised countries the share of manufacturing in the gross domestic product is far less than the share of the services sector. It seems important therefore to enlarge the scope to industries producing services or, more general, immaterial goods. One could, for example, investigate whether technological regime conditions also apply to patterns of innovative activities in services industries. Further, one can look for typical life cycle phenomena (such as shakeouts) in these sectors. Finally, one could elaborate on the innovative processes in services industries. Many services are characterised by the phenomenon that the supplier and the demander have to meet physically in order to perform a transaction. Hence, for this type of services production and consumption takes place simultaneously. Is it possible then to distinguish product innovations from process innovations?

Empirical research on services industries, based on longitudinal firm-level data, may answer some interesting questions as well. For instance, do the empirical regularities observed in manufacturing industries also emerge in services industries? An answer to this question should illuminate additional similarities and differences between manufacturing and services industries. Furthermore, one could investigate whether and explain why services industries differ with regard to their structural and dynamic properties. Whether such an investigation would resemble the present book remains an interesting question.

REFERENCES

Abernathy, W. and Utterback, J. (1978): "Patterns of Industrial Innovation". *Technology Review* 80: 41–47.

Audretsch, D. (1987): "An Empirical Test of the Industry Life Cycle". *Weltwirtschaftliches Archiv* 123 (2): 297–308.

Audretsch, D. (1997), "Technological Regimes, Industrial Demography and the Evolution of Industrial Structures". *Industrial and Corporate Change* 6 (1): 49–82.

Bain, J. (1951): "Relation of Profit Rate to Industrial Concentration: American Manufacturing 1936–40". *Quarterly Journal of Economics* 65: 293–324.

Bain, J. (1954): "Economies of Scale, Concentration, and the Condition of Entry in Twenty Manufacturing Industries". *American Economic Review* 44: 15–39.

Bain, J. (1956): "Barriers to New Competition". Cambridge, MA: Harvard University Press.

Baldwin, J. (1995): "The Dynamics of Industrial Competition: A North American Perspective". Cambridge: Cambridge University Press.

Baldwin, J. and Caves, R. (1998): "International Competition and Industrial Performance: Allocative Efficiency, Productive Efficiency, and Turbulence". In: "The Economics and Politics of International Trade". Gary Cook, (ed.) London: Routledge: 57–84.

Baldwin, J. and Rafiquzzaman, M. (1995): "Restructuring in the Canadian Manufacturing Sector from 1970 to 1990: Industry and Regional Dimensions of Job Turnover". *Research Paper Series* 78, Analytical Studies Branch, Statistics Canada.

Baumol, W., Panzar, J. and Willig, R. (1982): "Contestable Markets and the Theory of Industry Structure". New York: Harcourt Brace Jovanovich.

Beggs, A. and Klemperer, P. (1992): "Multi-Period Competition with Switching Costs". *Econometrica* 60: 651–666.

Breschi, S., Malerba, F. and Orsenigo, L. (1996): "Technological Regimes and Schumpeterian Patterns of Innovation". Paper prepared for the Meeting of the Josef A. Schumpeter Society, Stockholm, June 1996.

Bork, R. (1978): "The Antitrust Paradox". New York: Basic Books.

Cable, J. and Schwalbach, J. (1991): "International Comparisons of Entry and Exit". In: Geroski, P. and Schwalbach, J. (eds.): "Entry and Market Contestability". Oxford: Blackwell: 257–281.

Caves, R. (1998): "Industrial Organization and New Findings on the Turnover and Mobility of Firms". *Journal of Economic Literature* 36: 1947–1982.

Cefis, E. (1998): "Persistence in Profitability and in Innovative Activities". Paper presented at the European Meeting on Applied Evolutionary Economics held in Grenoble, June 1999.

Coombs, R. (1988): "Technological opportunities and industrial organisation". In: Dosi, G., C. Freeman, R. Nelson, G. Silverberg and L. Soete (eds.): "Technical Change and Economic Theory". London: Frances Pinter: 295–308.

Davies, S. (1979): "The Diffusion of Process Innovations". Cambridge: Cambridge University Press.

Dean, J. (1950): "Pricing Policies for New Products". *Harvard Business Review* 28: 45–53.

Demsetz, H. (1973): "Industry Structure, Market Rivalry, and Public Policy". *Journal of Law and Economics* 16: 1–9.

Dosi, G. (1982): "Technological Paradigms and Technological Trajectories". *Research Policy* 11: 147–162.

Dosi, G. (1988a): "Sources, Procedures and Microeconomic Effect of Innovation". *Journal of Economic Literature* 26: 1120–1171.

Dosi, G. (1988b): "The Nature of the Innovative Process". In: Dosi, G., C. Freeman, R. Nelson, G. Silverberg and L. Soete (eds.): "Technical Change and Economic Theory". London: Frances Pinter: 221–238.

Dosi, G., O Marsili, L. Orsenigo and R. Salvatore (1995): 'Learning, Market Selection and the Evolution of Industrial Structures'. *Small Business Economics* 7, 411–436.

Dosi, G., F. Malerba, O. Marsili and L. Orsenigo (1997): 'Industrial Structures and Dynamics: Evidence, Interpretations and Puzzles'. *Industrial and Corporate Change* 6 (1): 3–24.

Ekelund, R., and Hébert, R. (1990): "A History of Economic Theory and Method". Singapore: McGraw-Hill.

Ericson, R., and Pakes, A. (1995): "Markov-Perfect Industry Dynamics: A Framework for Empirical Work". *Review of Economic Studies* 62: 53–82.

Ericson, R., and Pakes, A. (1998): "Empirical Implications of Alternative Models of Firm Dynamics". *Journal of Economic Theory* 79: 1–45.

Evans, D. (1987a): "Tests of Alternative Theories of Firm Growth". *Journal of Political Economy* 95 (4): 657–674.

Evans, D. (1987b): "The Relationship between Firm Growth, Size, and Age: Estimates for 100 Manufacturing Industries". *Journal of Industrial Economics* 35 (4): 567–581.

Fisher, F. (1989): "Games economists play: a noncooperative view". *Rand Journal of Economics* 20 (1): 113–124.

Flaherty, M. (1980): "Industry Structure and Cost-Reducing Investment." *Econometrica* 48 (5): 1187–1209.

Freeman, C. (1974): "The Economics of Industrial Innovation". Harmondsworth: Penguin.

Freeman, C. (1982). "The Economics of Industrial Innovation". London: Frances Pinter.

Geroski, P. (1995): "What do we know about entry?" *International Journal of Industrial Organization* 13: 421–440.

Geroski, P. (1998): "The growth of firms in theory and in practice". Paper prepared for the 1998 DRUID Conference on "Competencies, Governance and Entrepreneurship".

Griliches, Z. (1958): "Research Costs and Social Returns: Hybrid Corn and Related Innovations". *Journal of Political Economy* 66: 419–431.

Gort, M. and Klepper, S. (1982): "Time paths in the diffusion of product innovations". *The Economic Journal* 92: 630–653.

Haltiwanger, J. (1997): "Measuring and Analyzing Aggregate Fluctuations: The Importance of Building from Microeconomic Evidence." *St. Louis Fed. Reserve Bank Econ. Review* 79 (3): 55–77.

Hannan, M. and Carroll, G. (1992): "Dynamics of organizational populations: density, legitimation and competition". New York: Oxford University Press.

Hannan, M. and Freeman, J. (1989): "Organizational ecology". Cambridge, MA: Harvard University Press.

Hayes, R. and Wheelwright, S. (1979a): "Link Manufacturing Process and Product Life Cycles". *Harvard Business Review* 57: 133–140.

Hayes, R. and Wheelwright, S. (1979b): "The Dynamics of Process-Product Life Cycles". *Harvard Business Review* 57: 127–136.

Hopenhayn, H. (1992): "Entry, Exit, and Firm Dynamics in the Long Run". *Econometrica* 60: 1127–1150.

Iwai, K. (1984a): "Schumpeterian Dynamics. An Evolutionary Model of Innovation and Imitation". *Journal of Economic Behavior and Organization* 5: 159–190.

Iwai, K. (1984b): "Schumpeterian Dynamics, Part II. Technological Progress, Firm Growth and 'Economic Selection'". *Journal of Economic Behavior and Organization* 5: 321–351.

Jensen, J., and McGuckin, R. (1997): "Firm performance and Evolution: Empirical Regularities in the US Microdata". *Industrial and Corporate Change* 6 (1): 25–47.

Jovanovic, B. (1982): "Selection and the evolution of industry". *Econometrica* 50 (3): 649–670.

Jovanovic, B. and MacDonald, G. (1994): "The Life Cycle of a Competitive Industry". *Journal of Political Economy* 102 (2): 322–347.

Kamien, M. and Schwartz, N. (1982): "Market Structure and Innovation". Cambridge: Cambridge University Press.

Kaniovski, Y. (1998): "Interdependent Search and Industry Dynamics: on Ericson and Pakes (1995)". IIASA Interim Report IR-98–031.

Klepper, S. (1996): "Entry, Exit, Growth, and Innovation over the Product Life Cycle". *American Economic Review* 86 (3): 562–583.

Klepper, S. (1997): "Industry Life Cycles". *Industrial and Corporate Change* 6 (1): 145–181.

Klepper, S. and Graddy, E. (1990): "The evolution of new industries and the determinants of market structure". *RAND Journal of Economics* 21 (1): 27–44.
Klepper, S. and Simons, K. (1999): "Industry Shakeouts and Technological Change". Paper presented at the European Meeting on Applied Evolutionary Economics held in Grenoble, June 1999.
Kuhn, T. (1962): "The Structure of Scientific Revolutions". Chicago: Chicago University Press.
Levin, R. et al. (1984): "Survey research on R&D appropriability and technological opportunity. Part 1: Appropriability". New Haven: Yale University Press.
Levitt, T. (1965): "Exploit the Product Life Cycle". Harvard Business Review 18: 81–94.
Lippman, S. and Rummelt, R. (1982): "Uncertain Immitability: An Analysis of Interfirm Differences Under Competition". *Bell Journal of Economics* 13: 418–438.
Marshall, A. (1936): "Principles of Economics" (8th edition). London: Macmillan.
Marsili, O. (1999): "The Anatomy and Evolution of Industries. Technological change and industrial dynamics". SPRU , University of Sussex (unpublished dissertation).
Malerba, F. and Orsenigo, L (1994): "The Dynamics and Evolution of Industries". IIASA Working Paper WP-94-120.
Malerba, F., and L. Orsenigo (1995): "Schumpeterian patterns of innovation". *Cambridge Journal of Economics* 19, 47–65.
Malerba, F., and L. Orsenigo (1996): 'Schumpeterian patterns of innovation are technology specific'. *Research Policy* 25 (3), 451–478.
Martin, S. (1994): "Industrial Economics. Economic Analysis and Public Policy". Prentice-Hall, New Jersey.
Maskin, E. and Tirole, J. (1988): "A Theory of Dynamic Oligopoly: I and II". *Econometrica* 56: 549–600.
Mason, E. (1957): "Economic Concentration and Monopoly Problem. Cambridge, MA: Harvard University Press.
Nelson, R. (1994): "What has been the Matter with Neoclassical Growth Theory?" In: Silverberg, G. and Soete, L. (eds.): "The Economics of Growth and Technical Change". Aldershot: Edward Elgar Publishing Limited.
Nelson, R. (1995): "Recent Evolutionary Theorizing About Economic Change". *Journal of Economic Literature* 13: 48–90.
Nelson, R. and Winter, S. (1977): "In search of a useful theory of innovation". *Research Policy* 6: 36–76.
Nelson, R. and Winter, S. (1982): "An Evolutionary Theory of Economic Change". Cambridge: Harvard University Press.
Neumann, J. Von, and Morgenstern, O. (1944): "The Theory of Games and Economic Behavior". New York: Wiley.
Orsenigo, L. (1989): "The Emergence of Biotechnology: Institutions and Markets in Industrial Innovation". London: Pinter Publishers.
Rasmussen, E. (1989): "Games and Information: an introduction to game theory". Oxford: Basil Blackwell.
Reinganum, J. (1981): "On the Diffusion of New Technology: a game theoretic approach". *Review of Economic Studies* 48: 395–405.
Rosenberg, N. (1976): "Perspectives on technology". Cambridge: Cambridge University Press.
Rosenthal, R. and Spady, R. (1989): "Duopoly with both Ruin and Entry". *Canadian Journal of Economics* 22: 834–851.

Sahal, D. (1985): "Technology Guide-Posts and Innovation Avenues". *Research Policy* 14 (2): 61–82.

Saviotti, P. and Metcalfe, S. (1984): "A Theoretical Approach to the Construction of Technological Output Indicators". *Research Policy* 13 (3): 141–151.

Scherer, F. (1967): "Market Structure and the Employment of Scientists and Engineers". *American Economic Review* 57: 524–531.

Scherer, F. (1980): "Industrial Market Structure and Economic Performance" (2nd edition). Boston: Houghton Mifflin Company.

Scherer, F. (1986): "Innovation and Growth. Schumpeterian Perspectives". Cambridge: MIT Press.

Schmalensee, R. (1989): "Inter-Industry Studies of Structure and Performance". In: Schmalensee, R. and Willig, R., (eds.): "Handbook of Industrial Organization, Volume II". Amsterdam: North-Holland Publishing Co.: 951–1009.

Schmookler, J. (1966): "Invention and Economic Growth". Cambridge: Harvard University Press.

Schumpeter, J. (1912), "The Theory of Economic Development". Oxford: Oxford University Press.

Schumpeter J. (1939): "Business Cycles: a Theoretical, Historical and Statistical Analysis of the Capitalist Process". New York: McGraw-Hill.

Schumpeter, J. (1942), "Capitalism, Socialism and Democracy". New York: Harper and Brothers.

Shy, O. (1996): "Technology revolutions in the presence of network externalities". *International Journal of Industrial Organization* 14: 785–800.

Silverberg, G., and Lehnert, D. (1993): "Long waves and 'evolutionary chaos' in a simple Schumpeterian model of embodied technical change". *Structural Change and Economic Dynamics* 4: 9–37.

Silverberg, G., and Verspagen, B. (1994a): "Collective learning, innovation and growth in a boundedly rational, evolutionary world". *Journal of Evolutionary Economics* 4: 207–226.

Silverberg, G., and Verspagen, B. (1994b): "Learning, innovation and economic growth: a long-run model of industrial dynamics". *Industrial and Corporate Change* 3: 199–223.

Simon, H. (1959): "Theories of Decision Making in Economics". *American Economic Review* 49: 253–283.

Simon, H. (1965): "Administrative Behavior". New York: Free Press.

Stoneman, P. (1991): "Technological Diffusion: The Viewpoint of Economic Theory". In: Mathias, P. and Davis, J. (eds.): "Innovation and technology in Europe: From the eighteenth century to the present day. The Nature of Industrialization series". Oxford and Cambridge: Blackwell: 162–184.

Suárez, F. and Utterback, J. (1995): "Dominant Designs and the Survival of Firms". *Strategic Management Journal* 16: 415–430.

Sutton, J. (1991): "Sunk Costs and Market Structure". Cambridge, MA: MIT Press.

Sutton, J. (1997): "Gibrat's Legacy". *Journal of Economic Literature* 35: 40–59.

Sutton, J. (1998): "Technology and Market Structure". Cambridge, MA: MIT Press.

Teece, D. (1982): "Toward an Economic Theory of the Multiproduct Firms". *Journal of Economic Behaviour and Organisation* 3 (1): 39–63.

Teece, D. (1986): "Profiting from Technological Innovation". *Research Policy* 15 (6): 285–306.

Utterback, J. and Abernathy, W. (1975): "A Dynamic Model of Process and Product Innovation". *Omega* 3 (6): 639–656.

Vernon, R. (1966): "International Investment and International Trade in the Product Life Cycle". *Quarterly Journal of Economics* 80: 190–207.
Vernon, R. (1979): "The Product Life Cycle Hypothesis in a New International Environment". *Oxford Bulletin of Economics and Statistics* 41: 255–267.
Weiss, L. (1974): "The Concentration-Profits Relationship and Antitrust". In: Goldschmid, H., Mann, H. and Weston, J. (eds.): "Industrial Concentration: The New Learning". Boston: Little Brown.
Windrum, P. and Birchenhall, C. (1998): "Is Life Cycle Theory a Special Case? Dominant Designs and the Emergence of Market Niches through Coevolutionary Learning". *Structural Change and Economic Dynamics* 9: 109–134.
Winter, S. (1984): 'Schumpeterian Competition in Alternative Technological Regimes'. *Journal of Economic Behavior and Organization* 5: 287–320.

This research has been executed at the Center for *R*esearch economic Microdata (Ce*Re*M) of Statistics Netherlands. The views expressed in this research are those of the author and do not necessarily reflect the policies of Statistics Netherlands.

SUBJECT INDEX